When you drink water,
think of the source!

Chinese proverb

CO_2 offset

For every printed book sold, a CO_2 offset is created
by supporting projects to reduce CO_2 emissions
as well as through tree planting and forest protection

Christian Becke

Carbon Dioxide

CO_2

Greenhouse Gas in the Air

and

Carbonic Acid System in Water

FSC
www.fsc.org
MIX
Papier aus verantwortungsvollen Quellen
Paper from responsible sources
FSC® C105338

Bibliographic Information from the German National Library: The German National Library lists this publication in the German National Bibliography; detailed bibliographic data is available online at dnb.dnb.de.

Translation: tolingo
Reprint July 2026 - with only minor changes

Publisher: BoD · Books on Demand GmbH, Überseering 33, 22297 Hamburg, bod@bod.de
Print: Libri Plureos GmbH, Friedensallee 273, 22763 Hamburg

ISBN: 978-3-6963-9289-5

Foreword

Carbon dioxide (CO_2) is an important chemical compound for the earth and life on earth. Since CO_2 plays a major role in the two spheres of air and water and these CO_2 occurrences are in exchange with each other, these spheres cannot be considered separately.

The previous version of this book, "Carbonic acid in Water", was still primarily tailored to CO_2 and carbon dioxide in the aqueous phase. The contents of this book have been revised and a further focus placed on CO_2 as a greenhouse gas in the atmosphere, which is also reflected in the new title.

Since the subject area is very extensive, not all facets can be considered. The fundamentals of the systems are dealt with here and calculation possibilities in the aqueous system are presented, which are deepened via calculation examples.

The example calculations in this book are generally performed for 420 ppmv CO_2 in the gas phase, a value reached in the atmosphere in April 2022. However, due to the increasing value in the atmosphere, the calculations must always be updated.

The calculations in the carbonic acid system are very complex with high required accuracy. Each user must decide for himself what accuracy and scope he needs.

The standards and limit values cited in this book always refer to the time the text was written and should always be checked and verified to ensure they are up to date.

Radolfzell at Lake Constance, July 2026

Christian Becke

Contents

1 Properties of carbon dioxide

Equivalent designations:

Carbon dioxide	Carbon(IV) oxide
Dioxomethane	Carbonic acid gas

= Anhydride of carbonic acid

Anhydride of carbonic acid means
CO_2 is formed when a water molecule splits off from carbonic acid:

$$H_2CO_3 \rightarrow CO_2 + H_2O$$

Conversely, CO_2 forms carbonic acid with water:

$$CO_2 + H_2O \rightarrow H_2CO_3$$

1.1 General data

Carbon dioxide is a chemical compound consisting of one carbon atom and two oxygen atoms. CO_2 belongs to the group of inorganic compounds.

The carbon in CO_2 has the oxidation number +IV. CO_2 is thus the most oxidized form of carbon C. The organic compound methane CH_4 (-IV) is the most reduced.

Oxidation number of carbon C -IV to +IV

Carbon dioxide is a colorless, odorless and non-flammable gas that occurs in the atmosphere and dissolved in water. In bound mineral form it is found as carbonate rock (e.g. lime) in the lithosphere (earth's crust and upper mantle).

The compound of CO_2 is very stable and decomposes under atmospheric pressure only at very high temperatures to carbon monoxide and oxygen (51.7% at 2,606 °C) and at even higher temperatures to carbon and oxygen [1].

$$CO_2 + 67{,}635\ \text{kcal} \leftrightarrows CO + \tfrac{1}{2}O_2 \qquad [1]$$

$$CO_2 + 94{,}051\ \text{kcal} \leftrightarrows C + O_2 \qquad [1]$$

From the state diagram for CO_2 (Figure 1.1-1:) it can be seen that under the pressure and temperature conditions on earth carbon dioxide can naturally exist neither as a liquid nor as a solid but only as a gas. Due to the temperature conditions on Mars, solid CO_2 is present there at the polar caps, for example.

However, carbon dioxide can be technically liquefied or converted into solid form ("dry ice"). Solid CO_2 has no melting point at normal pressure, so it does not become liquid when the temperature rises, but sublimates at -78.48 °C and passes directly into the gas phase. Liquid CO_2 can be produced under pressure at temperatures below 31 °C (critical temperature) and filled into gas cylinders. The required pressure in the temperature range from -56.6 to 31.0 °C is at least 5.19 bar (for example 34.3 bar at 0 °C or 19.3 bar at -20 °C) [1].

The critical point of CO_2 is reached at 73.83 bar and 31.0 °C. In the supercritical range, the gas phase and the liquid phase are indistinguishable.

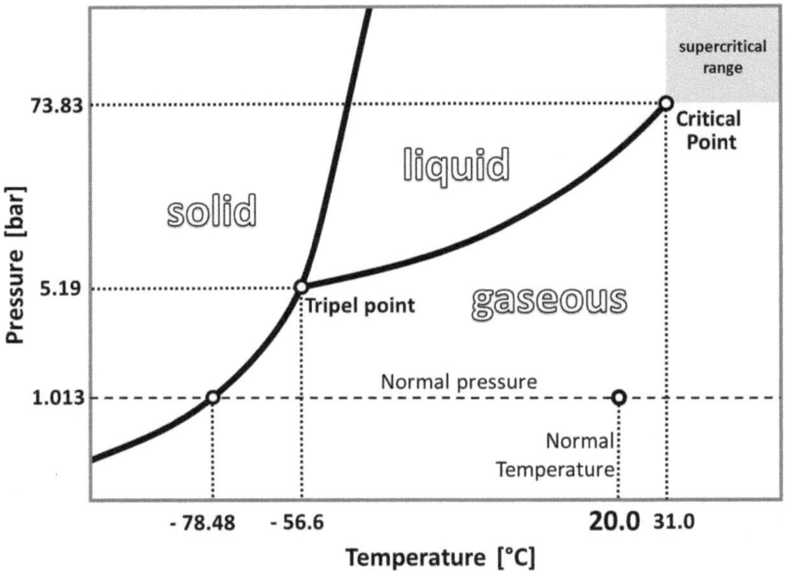

Figure 1.1-1: State diagram (phase diagram) for CO_2 (schematic, course of curve qualitative, ordinate logarithmic) (1 bar = 100 kPa)

For content data of the gas in the gas phase, either the unit ppmv (parts per million by volume) or vol% is used:

$$1 \text{ ppmv } (\text{mL/m}^3) = 0.0001 \text{ vol\%} \quad \text{or} \quad 1 \text{ vol\%} = 10{,}000 \text{ ppmv}$$

General data for CO_2 are shown in Table 1.1-1.

Table 1.1-1: Carbon Dioxide Data - Overview

CAS number	124-38-9		[3]
EC number	204-696-9		[3]
Food additive	E290		[4]
Sum formula	CO_2		[1]
Components	Carbon C Oxygen O		[1]
Oxidation number	+IV		[1]
ΔH	-393.5	kJ/mol	[1]
Molar mass	44.0098	g/mol	[1]
Density	1.9768	kg/m³ 0 °C, 1013 hPa	[1]
Density of air	1.2920	kg/m³ 0 °C, 1013 hPa	[3]
Density ratio to air	1.53		[3]
Density at -79 °C (solid)	1.56	kg/L -79 °C 1013 hPa	[5]
State of aggregation	gaseous		[1]
Triple point	-56.6 5.19	°C bar	[3]
Sublimation point	- 78.48	°C	[1]
Critical temperature	31.00	°C	[1]
Critical pressure	73.83	bar	[3]
Critical density	0.468	g/cm³	[3]
HENRY constant – Freshwater	38.674	mol/(m³·bar) , 20 °C	[6]
HENRY constant – Salt water	32.571	mol/(m³·bar) , 20 °C	[6]
AGW (workplace)	0.5	vol%	[7]
AGW (workplace)	9100	mg/m³	[7]
Global **W**arming **P**otential	1		[3]
Ozone Depletion Potential	0		[3]

In the atmosphere, CO_2 is equally distributed alongside nitrogen and oxygen and the other trace gases. If it occurs somewhere in a concentrated form, e.g. from crevices in the ground (volcanism), in caves (e.g. Cave of Dogs near Naples), in well shafts or even in fermentation cellars, it accumulates at the bottom, since it is significantly specifically heavier than air (factor 1.53). Therefore, there is a great potential for danger when entering such places, even at normal oxygen concentration (21%).

At levels of 5% or more of carbon dioxide in the inhaled air, headaches and dizziness occur; at higher concentrations, shortness of breath and unconsciousness occur. Carbon dioxide concentrations of 8% cause death within a short time. [2]

The occupational exposure limit value (In Germany: AGW, formerly MAK value) for CO_2 is 5,000 ppmv = 0.5% by volume. According to the Occupational Health and Safety Ordinance, the occupational exposure limit indicates the concentration of a substance up to which acute or chronic adverse health effects are generally not to be expected in the workplace [8]. This is based on the assumption that exposure usually lasts 8 hours and that the average working week is 40 hours.

The Ozone Depletion Potential (ODP) of carbon dioxide is 0, meaning that the CO_2 content of the atmosphere has no effect on the depletion of the ozone layer [9].

Global warming potential is discussed in 2.2.

1.2 Molecular structure

Carbon dioxide

The CO_2 molecule has a linear structure.

Sum formula: CO_2

Structural formula: O=C=O

Electron formula:
(with mesomeric shapes)

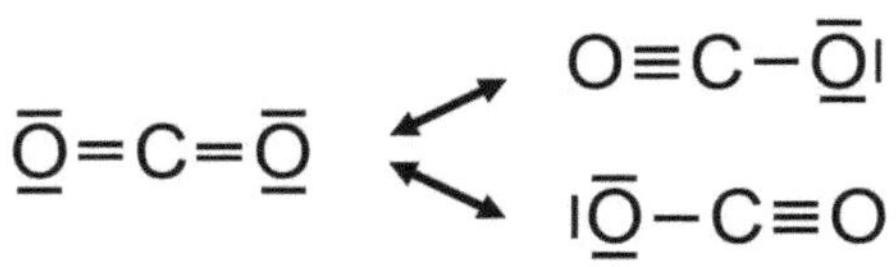

Due to the higher electronegativity of oxygen of 3.5 compared to carbon with 2.5 (values according to [10]), partial charges are formed in the CO_2 molecule. Due to the linear and symmetrical structure, the molecule is neutral to the outside despite the polar bond and is not a permanent dipole.

$$\overset{\delta-}{O}=\overset{\delta+}{C}=\overset{\delta-}{O}$$

Due to the oscillations of the electrons within the molecule, however, an uneven charge distribution occurs at times, so that a dipole character then exists (fluctuating dipole). This can lead to interactions with other molecules.

Carbonic acid

Sum formula:	H_2CO_3	
Diprotic acid	→ 2 dissociation stages	(Chapter 4.3)
	→ 2 pK values	(Chapter 4.3)

Carbonic acid is not presentable in pure form. Their salts can be isolated and are partly present as solids:

Hydrogen carbonates and carbonates = salts of carbonic acid

The carbonate ion has a planar structure with OCO angles of 120° [1].

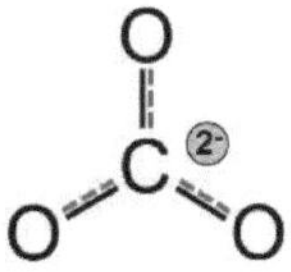

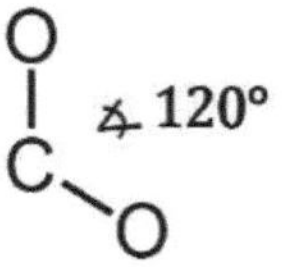

Mesomeric shapes of the carbonate ion:

Structural formula:

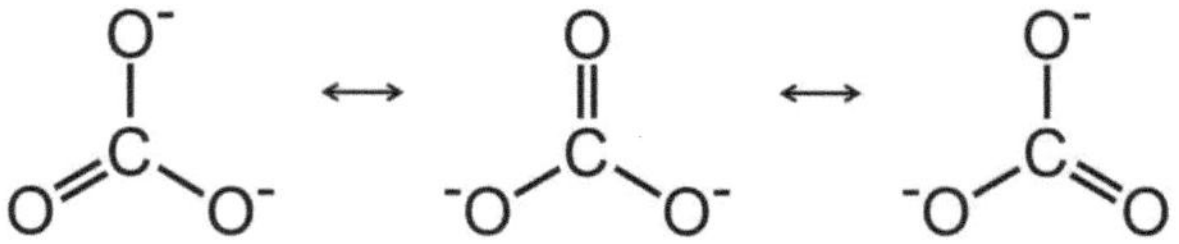

Electron formula:

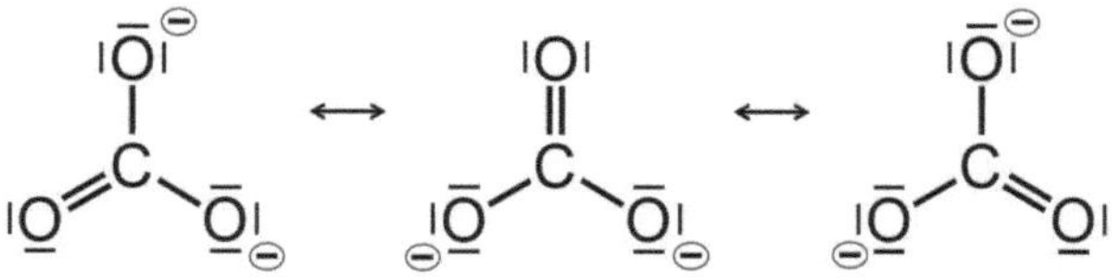

1.3 Radiation absorption

Electromagnetic radiation can cause interactions with molecules in the infrared (IR) range due to molecular vibrations and rotational vibrations of gaseous molecules.

The infrared range includes radiation of wavelength from 780 nm to 1000 µm and is limited downward (shorter wavelength) by the visible range of radiation and upward (longer wavelength) by the range of microwaves (Table 1.3-1).

Table 1.3-1: Classification of the wavelength ranges [µm] of the infrared radiation according to DIN 5031-7 [11]

<table>
<tr><td></td><td>Visible light</td><td></td><td></td><td></td><td></td></tr>
<tr><td rowspan="4">0.78 µm

1000 µm</td><td rowspan="4">Infrared radiation</td><td>0.78-1.4</td><td>IR-A</td><td rowspan="2">Near infrared</td><td rowspan="2">NIR</td></tr>
<tr><td>1.4-3.0</td><td>IR-B</td></tr>
<tr><td>3.0-50</td><td rowspan="2">IR-C</td><td>Medium infrared</td><td>MIR</td></tr>
<tr><td>50-1000</td><td>Far infrared</td><td>FIR</td></tr>
<tr><td></td><td>Microwaves</td><td></td><td></td><td></td><td></td></tr>
</table>

When vibrations are excited in the region of molecular bonds, radiation is absorbed, converted into vibrational energy and re-emitted in an undirected manner. The emitted frequency is lower or higher than the absorbed frequency.

Vibrational frequency of molecules is specific to functional groups of molecules and depends on bond strength of chemical bond and vibrational masses. [1]

The CO_2 molecule has a linear structure and is 3-atomic. Therefore it has 3 degrees of freedom and 3N-5 oscillations can occur. With the number of atoms N=3 we get (3x3)-5=4 oscillations [12]. One symmetrical and one asymmetrical valence oscillation as well as two deformation oscillations are possible (Table 1.3-2).

For a molecule to interact with electromagnetic radiation (IR-active), a dipole moment or an inducible dipole moment must be present.

$$\overset{\delta-}{O}=\overset{\delta+}{C}=\overset{\delta-}{O}$$

The CO_2 molecule has no dipole moment in the ground state, but an induced dipole moment is possible if the partial charges are asymmetrically displaced by the vibrations (see also chapter 1.2). If the oscillations occur symmetrically with respect to the center of symmetry (b in Table 1.3-2), no dipole moment is generated, that is, no IR radiation is absorbed (IR inactive). [13] [14]

Table 1.3-2: Possible oscillations of the CO_2 molecule [13] [14]

a	b	c	d
δ- δ+ δ- O=C=O ←← →←	δ- δ+ δ- O=C=O ←← →→	δ- δ+ δ- O=C=O ↓ ↑ ↓	δ- δ+ δ- O=C=O + - +
asymmetric	symmetrical	in the image plane	perpendicular to the image plane
Valence oscillation		Deformation vibration	
IR active	--	IR active	IR active
$\tilde{\nu}$ = 2349 cm^{-1}	$\tilde{\nu}$ = 1338cm^{-1}	$\tilde{\nu}$ = 667 cm^{-1}	$\tilde{\nu}$ = 667 cm^{-1}
λ = 4.26 µm	λ = 7.47 µm	λ = 14.99 µm	λ = 14.99 µm

The oscillations are characterized by the wavelength λ or the wavenumber $\tilde{\nu}$ of the radiation. The wave number indicates the number of oscillations per unit length.

Relationship between wavenumber and wavelength:

$$\tilde{\nu} = 1/\lambda \qquad (\text{cm}^{-1})$$

$\tilde{\nu}$ Wave number
λ Wavelength

The two deformation oscillations (c and d in Table 1.3-2), which differ only in the plane, have the same wavenumber.

Thus, the maxima of absorption of IR radiation by the CO_2 molecule are at λ = 4.26 µm and λ = 14.99 µm, respectively, and at the wavenumbers $\tilde{\nu}$ = 2349 cm^{-1} and $\tilde{\nu}$ = 667 cm-^{1}. Interactions with the rotation of the gaseous molecules give rise to vibrational bands in the IR spectrum consisting of a larger number of closely spaced absorption lines (Figure 1.3-1).

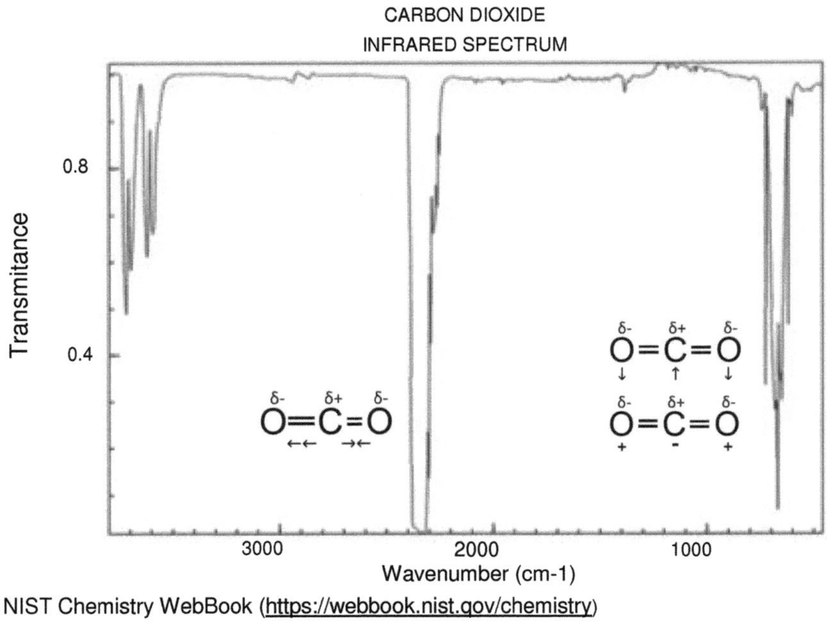

Figure 1.3-1: IR spectrum of CO_2 (Graphic: NIST [15], supplemented)

2 The importance of carbon dioxide

Carbon dioxide CO_2

- is building block for life through conversion into organic matter
- is the main nutrient for plants, both on land and in water
- is part of the carbon cycle
- influences the climate as a greenhouse gas
- is involved in the earth's crust in the form of carbonates as mineral formers
- promotes weathering as a water constituent
- significantly influences the properties of a water
- is often a problem for drinking water quality
- often makes water treatment necessary
- is the engine for cold water geysers
- can cause death at high concentrations in the air

2.1 Carbon dioxide as the basis of life

The inorganic CO_2 molecule is the basis for building organic matter (assimilation) and thus the link between organic and inorganic matter (Figure 2.1-1).

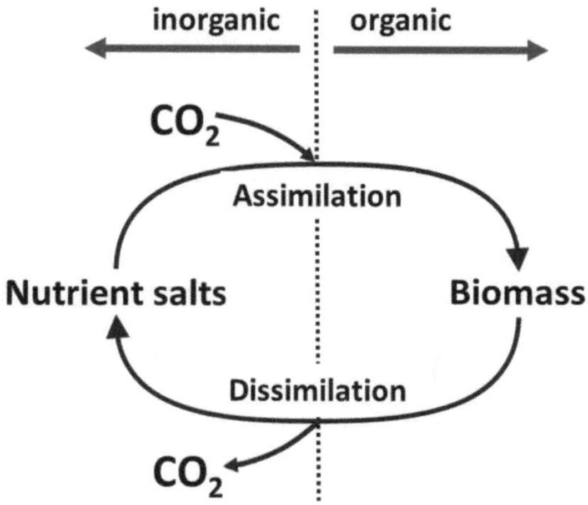

Figure 2.1-1: Carbon dioxide as a link between inorganic and organic matter

Organic compounds, i.e. the building materials of life, are based on reduced carbon compounds (oxidation number < IV) (organic chemistry = chemistry of carbon).

The CO_2 molecule is the most oxidized form of carbon (oxidation number +IV) and is reduced to hydrocarbon (organic) during assimilation by autotrophic organisms using energy.

*Autotrophic: Autotrophic organisms can produce energy through photosynthesis or chemosynthesis itself.

*Heterotrophic: Heterotrophic organisms are dependent on the supply of organic matter through the decomposition of which they gain energy.

The first organisms on earth did this chemoautotrophically (chemosynthesis). That is, they used the energy released in chemical oxidation reactions (chemical energy). The development of photosynthesis made it possible for other organisms to use light energy for this purpose (photoautotrophic). Photosynthesis enriched the atmosphere with oxygen (Figure 2.1-3).

The electrons required for reduction come from the oxidation of reduced compounds (e.g. $Fe^{2+} \rightarrow Fe^{3+}$) in chemosynthesis (lithotrophic) and from the oxidation of oxygen in the water molecule to O_2 in photosynthesis (phototrophic).

$$\underset{\substack{\textbf{oxidized}\\ \textbf{low energy}}}{CO_2} + \text{electrons } (e^-) + \textbf{energy} \rightarrow \underset{\substack{\textbf{reduced}\\ \textbf{high energy}}}{\textbf{organic compound}}$$

Figure 2.1-2: Conversion of CO_2 into organic matter (assimilation)

Plant growth and the production of plant biomass - the so-called primary production - is therefore, in addition to the need for mineral nutrients, primarily tied to the presence of CO_2 as a carbon source for building up organic matter (CO_2 as a nutrient).

Land plants use carbon dioxide from the air, while aquatic plants use CO_2 dissolved in water or hydrogen carbonate ions, depending on their adaptation (see 5.6).

In the primary processes of photosynthesis, plants gain energy from solar radiation, which they use to reduce CO_2 in the secondary processes and convert it into organic matter (assimilation).

$$6\, CO_2 + 12\, H_2\mathbf{O} \xrightarrow[\text{Solar energy}]{} \underset{\text{Glucose}}{C_6H_{12}O_6} + 6\, H_2O + 6\, \mathbf{O}_2$$

Figure 2.1-3: Sum formula of photosynthesis

The different labeling of the oxygen atoms in Figure 2.1-3 illustrates that the oxygen released during photosynthesis is split off from the water, while the oxygen from the CO_2 is contained in the glucose and in the water that is released.

As a rule, the CO_2 is bound in the so-called CALVIN cycle and a C-3 body with 3 carbon atoms (3-phosphoglyceric acid) is formed. Plants that assimilate CO_2 by this mechanism are called C-3 plants.

However, in other plants, the CO_2 is primarily bound to a C-3 body (phosphoenolpyruvate), creating a C-4 body, and only then transferred to the

CALVIN cycle. Plants in which, according to this mechanism, a C-4 body is the first substance in which the bound CO_2 is contained are called C-4 plants.

The C-4 plants perform photosynthesis more effectively and outperform the C-3 plants with a higher growth rate. When used to produce bioenergy, such crops are preferred (e.g. corn). For example, reed grass would also be suitable [16].

Thus, all organic compounds are reduced with respect to CO_2 and thus are energy-rich compounds. Therefore, you can get energy from them. Many living things do this by taking up organic compounds and oxidizing them to CO_2. They use the energy for their own metabolism and to build up their own biomass (heterotrophic lifestyle). This also includes humans, who release the decomposition product CO_2 through their lungs. Biochemical CO_2 splitting takes place predominantly in the so-called citric acid cycle. The final energy gain then occurs in the respiratory chain.

organic compound	→	CO_2 + electrons (e^-) + **energy**
reduced		**oxidized**
high energy		**low energy**

Figure 2.1-4: Energy gain through decomposition of organic matter

In Figure 2.1-5, the linkage of autotrophic and heterotrophic lifestyles is summarized. CO_2 is the starting product and the end product.

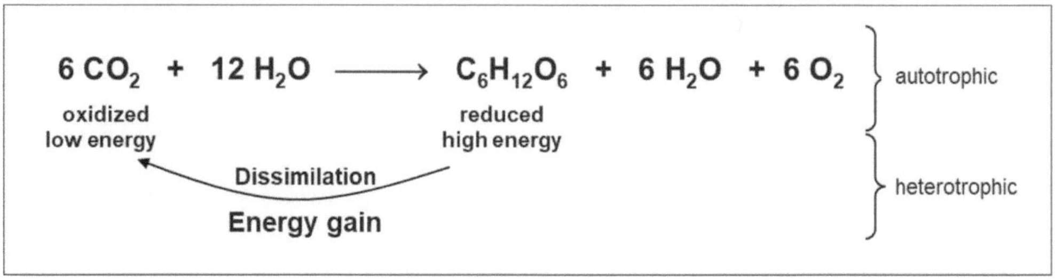

Figure 2.1-5: Linking autotrophic and heterotrophic lifestyles

Humans also use fossil organic energy sources (petroleum, coal, natural gas (methane)) for industrial energy production or as fuel and burn (oxidize) them with the end product CO_2.

2.2 CO_2 as a greenhouse gas

Greenhouse gases in the atmosphere absorb certain portions of the radiated by the earth heat radiation (Infrared radiation) and re-emit heat radiation in all directions. This results in a proportionate re-radiation effect in the direction of the earth, which leads to an increase in the temperature on the earth.

The earth's climate is influenced by this so-called greenhouse effect . The gases act like a greenhouse that does not allow some of the radiated heat to escape.

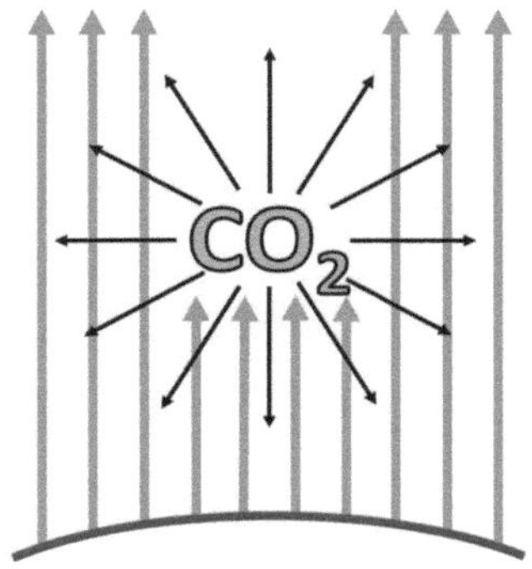

Figure 2.2-1: Schematic representation of the absorption and undirected radiation of heat radiation by CO_2 in the earth's atmosphere

The gas molecules must exhibit a dipole effect or an inducible dipole effect (chapter 1.3) in order to absorb infrared radiation.

The most potent natural greenhouse gas in the atmosphere is water vapor (gaseous water). However, the influence of humans on the water content of the atmosphere is small. Therefore, the focus of climate change is on gases that are strongly influenced in their concentration by humans.

The Kyoto Protocol (UN 2005) identifies six greenhouse gases whose emissions must be regulated, with an addition of 2015 [17] (Table 2.2-1:).

Table 2.2-1: Greenhouse gases according to Kyoto Protocol (UN 2005)

Natural gases	Fluorinated greenhouse gases (F-gases)
Carbon dioxide CO_2	hydrofluorocarbons containing hydrogen (HFC)
Methane CH_4	perfluorinated hydrocarbons (PFCs)
Nitrous oxide N_2O	Sulfur hexafluoride SF_6
	Nitrogen trifluoride NF_3 (from 2015 additionally)

The greenhouse potential of gases is very different. To make comparisons possible, one sets the Global Warming Potential GWP of CO_2 as a reference point with a value of 1 by definition.

All other climate-impacting gases are evaluated relative to the global warming potential of CO_2. They therefore indicate how high the influence of the gas in question on the greenhouse effect is compared to CO_2. The basis of the comparison is the mass of gases released in t.

These so-called CO_2 equivalents (CO_2eq, also CO_2e) can then be used to compare the effect of different gases on climate change.

The gases have a different residence time in the atmosphere. Once emitted, CO_2 is not degraded there and remains in the atmosphere for a long time. The concentration of carbon dioxide in the atmosphere is regulated by the very sluggish carbon cycle (chap. 2.4 and 3.1.2).

In contrast, methane (CH_4) is degraded in the atmosphere over time to carbon dioxide and water vapor. The lifetime of methane in the atmosphere is about 12.4 years [18]. The originally high greenhouse effect of methane is then replaced by the weaker effect of the carbon dioxide produced during decomposition.

$$CH_4 + 2\,O_2 \rightarrow CO_2 + 2\,H_2O \qquad \text{(Eq 2.2-1)}$$

Due to the different residence times of the gases in the atmosphere, this results in different values for the global warming potential depending on the period under consideration of 20 or 100 years (Table 2.2-2).

For the residence time in the atmosphere and for the global warming potentials GWP, varying values are found, depending on the source used. The values presented in Table 2.2-2 are from the 2013 Working Group I publication for the IPCC Fifth Assessment Report [19].

Values for greenhouse gases not listed here can also be found there.

Table 2.2-2: Global Warming Potentials (IPCC Fifth Assessment Report 2013) [19]

Gas	Retention period years	Global Warming Potential GWP over 20 years	100 years
CO_2	~	1	1
CH_4	12.4	84	28
N_2O	121	264	265

From the values in Table 2.2-2 it can be deduced that, when considered over 100 years, one ton of methane released contributes 28 times as much to the greenhouse effect as one ton of CO_2. For nitrous oxide (N_2O), the effect is 265 times stronger.

Since methane is degraded in the atmosphere and has a mean residence time of only 12.4 years, the initial effect on climate is much higher immediately after emission and decreases over time. For short-term time periods, therefore, the impact of methane emissions on climate is much greater than the 20- or 100-year values reflect.

In addition to the Global Warming Potential GWP, there is also the Global Temperature Potential GTP, which assesses the impact of emissions on the development of temperature, in each case again relative to the effect of CO_2 [20] [19].

Table 2.2-3: Temperature Potentials (IPCC Fifth Assessment Report 2013) [19]

Gas	Retention period Years	Temperature potential GTP over		
		20 years	50 years	100 years
CO_2	~	1	1	1
CH_4	12.4	67	14	4
N_2O	121	277	282	234

In Germany, according to the Federal Environment Agency (figures for 2021 [21]), 88.61 percent of the release of greenhouse gases (CO_2-equivalent CO_2eq) is accounted for by carbon dioxide, 6.28 percent by methane, 3.65 percent by nitrous oxide, and 1.46 percent by the F-gases (Table 2.2-4).

Table 2.2-4: Shares of greenhouse gases released (CO_2eq) in Germany (2021) [21]

Greenhouse gas	CO_2eq %
Carbon dioxide	88.61
Methane	6.28
Nitrous oxide	3.65
F-Gases	1.46

At 88.6%, carbon dioxide thus accounts for the largest share of greenhouse gases emitted in Germany.

If we look at by category in Germany, the energy sector accounts for the largest quantities of carbon dioxide equivalents (CO_2eq), at 84.2% (Table 2.2-5:).

Table 2.2-5: Shares of CO_2eq emissions by category in Germany (2021) [21]

Category	CO_2eq %
Energy-related emissions	84.2
Industrial processes	7.5
Agriculture	7.2
Waste Management	1.1

2.3 The carbon footprint

The extent of CO_2 release due to human activity is often presented as a CO_2 footprint.

Figure 2.3-1: Graphical representation of the carbon footprint

The carbon footprint is a measure of how much CO_2 is released by a person due to their lifestyle or for the production of a product they use. In Figure 2.3-2 values of the Federal Environment Agency for the German average per person and year can be found. Also included here are the quantities of methane and nitrous oxide produced as CO_2 equivalents (CO_2eq).

The ecological CO_2 footprint in Germany averages 11.17 t CO_2eq per year (Figure 2.3-2, data Federal Environment Agency 2020 [22]), of which 18.7% is attributable to mobility and 24.5% to housing and electricity. One third of the carbon footprint is caused by other consumption.

Public emissions	0.86 t	7.7 %	
Nutrition	1.69 t	15.1 %	
Mobility	2.09 t	18.7 %	
Housing and electricity	2.74 t	24.5 %	**11.17 t**
Other consumption	3.79 t	33.9 %	

Figure 2.3-2: CO_2 footprint, German average, per person and year (CO_2eq). Data: Umweltbundesamt (German Federal Environmental Agency) 2020 [22]

If you want to calculate your personal carbon footprint, you can find corresponding footprint calculators on the Internet (e.g. Federal Environment Agency [22])

Table 2.3-1: CO_2 footprints (CO_2eq) of selected food products [23]

Food	kg CO_2eq per kg food
Beef (average)	13.6
Butter	9.0
Cheese (average)	5.7
Chicken (average)	5.5
Fish, aquaculture	5.1
Pork (average)	4.6
Olive oil, disposable glass bottle	3.2
Egg	3.0
Palm fat	2.9
Milk, ESL, whole milk, composite carton	1.4
Beer, 0.5 L tin can	1.0
Cucumber	0.4
Milk substitute, oat drink	0.3

Technical products can also be assigned a carbon footprint, the Product Carbon Footprint (PCF). This carbon dioxide product footprint represents the emissions of a product or service in terms of the relevant climate gases as CO_2 equivalents (CO_2eq) over its life cycle. This covers all process stages of a product, starting with the extraction and transport of raw materials, the development, manufacture and use of the product. The energy used in each case has a significant share in the CO_2 pollution, as long as it was obtained from fossil energy sources.

The basis for determining the carbon footprint is the DIN EN ISO 14067 guideline: Greenhouse gases - Carbon footprint of products - Requirements and guidelines for quantification [24].

In addition to avoiding and reducing greenhouse gas emissions in the manufacture of products, offsetting is possible through other measures that reduce greenhouse gas emissions elsewhere.

With regard to the impact of products and measures on the climate, the following terms are distinguished.

Climate neutrality
The climate is not influenced by processes, productions, actions. This can be achieved, for example, through the sole use of renewable energy or through compensatory measures.

Climate positive
Measures that have a positive impact on the climate and counteract global warming, e.g. CO_2 reduction.

Greenhouse (gas) neutrality
No greenhouse gases are emitted. Nevertheless, influences on the climate can take place. For example, the sealing of soils and thus interference with the runoff behavior of precipitation and evaporation, or, for example, changes in the reflection of thermal radiation from surfaces (albedo effect).

CO_2 neutrality
In the narrow sense, no CO_2 is emitted or the CO_2 emission is fully compensated. Other greenhouse gases or other effects on the climate are not taken into account.

2.4 The carbon balance of the earth

A distinction is made between exogenous and endogenous occurrences of carbon on earth. The exogenous occurrences are distributed among the atmosphere, hydrosphere, and biosphere, and the endogenous occurrences are distributed among the lithosphere and the Earth's mantle. Between these occurrences in the different spheres there is an exchange through transformation processes and cycles are created. A distinction is made between the geological (inorganic) Carbon cycle (Figure 3.1-3:), which has existed since the formation of the earth, and the organic cycle (Figure 3.1-4:), which began with the emergence of life on earth.

The renewal time of carbon in the cycles is about up to 10,000 years in the exogenous cycle, while millions of years are assumed for the endogenous cycle [25]. For the more short-term anthropogenically caused climate problem, however, it is primarily the shorter cycles in the exogenous area that have an impact and are therefore the focus of these considerations.

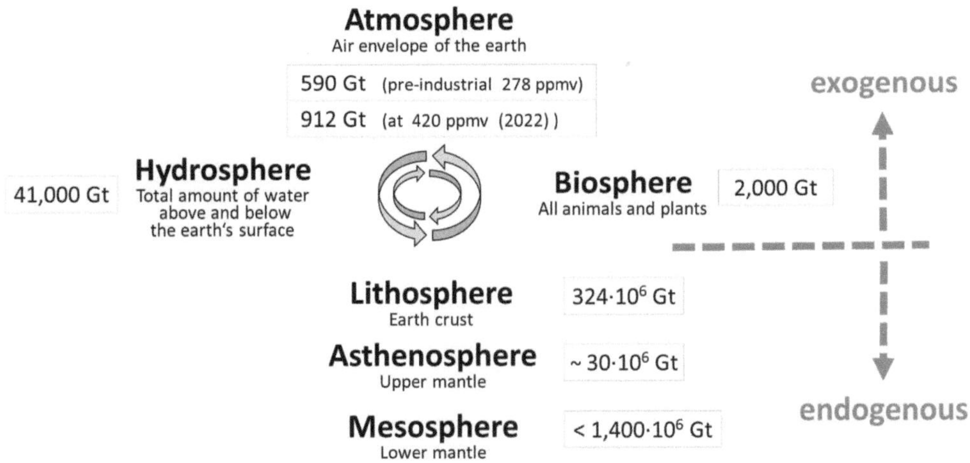

Figure 2.4-1: Carbon in the spheres of the earth
Data according to DCO [25] and [26]

In 2009, the large-scale scientific project Deep Carbon Observatory (DCO) [27] was founded with the aim of intensively researching carbon and exchange processes on earth. The DCO is an association of more than 1000 scientists. On the occasion of the 10th anniversary in 2019, a large number of publications on carbon balance were issued and the state of the research was published [28].

According to the results of the DCO project, the occurrence of carbon on earth comprises 1.85 billion Gt C. Of this, more than 99.99% is stored below the earth's surface (endogenously). Exogenously, there are just 43,500 Gt left (Figure 2.4-1) [25]. The values of DCO are given for pre-industrial times with an atmospheric value of 278 ppmv, which corresponds to 590 Gt in the atmosphere. There is about 912 Gt C in today's atmosphere at 420 ppmv (spring 2022 [29]).

(Example conversion from ppmv to Gt via the "homogeneous atmosphere" thought model [26] see chap. 10.1, calculation example 1)

The occurrences of elemental carbon (graphite, diamonds, fullerenes, and the rare modifications lonsdaleite and chaoite) play a minor role in the exogenous cycle in the carbon balance of the earth. Earth's known graphite reserves are about 325 million tons (2021) [30]. The other modifications of carbon are present in much smaller quantities. Carbon, therefore, in the exogenous cycle consists essentially of CO_2 and compounds derived from it. One could also speak of a CO_2 cycle.

There are different classifications for the classification of carbon compounds :

- Chemical
  - Inorganic

elementary	Graphite, diamonds, fullerenes, lonsdaleite and chaoite
gaseous	Carbon dioxide CO_2, carbon monoxide CO
bound	Carbonic acid, hydrogen carbonate, carbonate

  - Organic

gaseous	Methane CH_4 (e.g. natural gas), other organic gases
bound	animate nature, coal, petroleum, peat, lignite

- According to state

▸ gaseous	Carbon dioxide CO_2, carbon monoxide CO; methane CH_4, other organic trace gases
▸ bound/solid	Carbonic acid, hydrogen carbonate, carbonate, organic biomass, fossil deposits

3 Carbon dioxide content of air

3.1 Carbon dioxide in the atmosphere

At sea level, dry air consists of 99.95% nitrogen (N_2), oxygen (O_2) and the noble gas argon (Ar). Carbon dioxide is present at a level of 0.0418% by volume (418.56 ppmv, mean 2022 [31]). Other trace gases are found in the total with a proportion of 0.0028% by volume. These include methane (CH_4) and nitrous oxide (N_2O) as well as the other noble gases helium, neon, krypton and xenon (Table 3.1-1).

Table 3.1-1: Components of dry air in % by volume [32] [29]

78.08	vol%	Nitrogen	(N_2)	[32]
20.94	vol%	Oxygen	(O_2)	[32]
0.934	vol%	Argon	(Ar)	[32]
0.0418	vol%	Carbon dioxide	(CO_2)	[29]
0.0028	vol%	other trace gases		

1 ppmv (mL/m^3) = 0.0001 vol% or 1 vol% = 10,000 ppmv

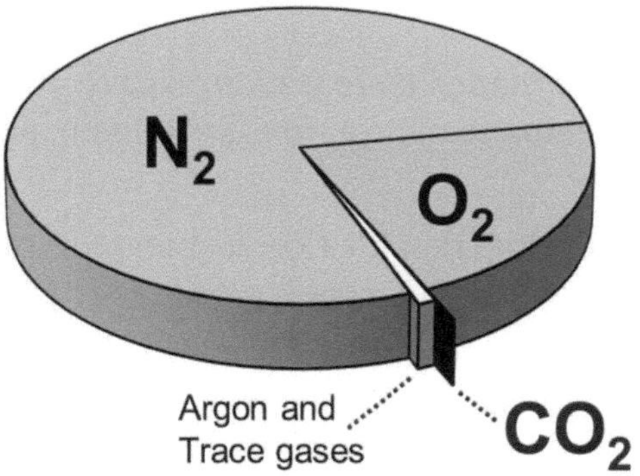

Figure 3.1-1: Dry air components

In natural air, water vapor (gaseous water*) is added at an average rate of 0.4% by volume [32].

(*the term water vapor refers to gaseous water, not condensed water, which is colloquially referred to as steam)

While the content of nitrogen and oxygen in the atmosphere is almost constant despite the participation in material cycles, especially the gases that are directly or indirectly influenced by humans fluctuate, especially CO_2, methane (CH_4, 1.895 ppmv 2021 [33]), nitrogen dioxide (NO_2, 0.334 ppmv 2021 [33]), and sulfur dioxide (SO_2) in their proportions in the air to some extent. The concentration of CO_2 has been increasing continuously since the beginning of industrialization and also shows seasonal fluctuations (chap. 3.1.5).

Carbon dioxide is one of the most important greenhouse gases, along with methane, nitrous oxide and water vapor. Methane is linked to CO_2 via the organic carbon cycle (chap. 3.1.2).

The graphical representation of the composition of the air (Figure 3.1-1) illustrates the small proportion of CO_2 in the atmosphere.

Despite this small proportion in the air, carbon dioxide has great importance in the atmosphere and in water.

3.1.1 Historical development of the CO_2 content of the atmosphere

According to current knowledge, the earth is about 4.5 billion years old [34]. The first gas shell was probably lost to space due to the still low gravity of the earth. The following primordial atmosphere consisted mainly of water vapor because of the very high temperatures of the earth in the beginning. This atmosphere was primarily still free of oxygen. At that time, oxygen was not present on earth in free gaseous form but was bound in oxides and was only later enriched in appreciable concentrations in the atmosphere by the photosynthesis of plants. However, recent studies show that O_2 may also have been formed in small amounts independently of biology by abiotic light-induced splitting of the CO_2 molecule into C and O_2 in the primordial atmosphere. [35]

It is assumed that the CO_2 content of the primordial atmosphere was very high. The earth's carbon balance argues for much higher concentrations of CO_2 in this atmosphere compared to today, before it was fixed in large quantities as carbonate in the earth's crust.

A high CO_2 content of the primordial atmosphere is also seen as a possible explanation for the "weak young sun paradox" described by SAGAN and MULLEN in 1972. It is about the contradiction between the low radiation power of the still young sun and the not correspondingly colder climate in the early history of the earth. This could have been caused by the climatic effect of a high CO_2 content of the atmosphere. [36] [37]

The following cooling of the earth led to a condensation of the water vapor and to the formation of the oceans (approx. 3.5 billion years ago). Rainfall removed a large fraction of CO_2 from the atmosphere and, dissolved in water, transferred it to the oceans, where it was fixed in significant quantities as precipitated carbonate in the earth's crust (carbon balance chap. 2.4). [26]

With the beginning of photosynthesis by organisms about 2.3 billion years ago, CO_2 was bound and converted into organic matter. In the process, gaseous oxygen was released from water molecules (chapter 2.1) and enriched in the atmosphere.

After that, two major cycles essentially determined the CO_2 content of the atmosphere. The inorganic (geologic) carbon cycle (Figure 3.1-3:) has been important from the beginning, except for the fraction of sedimented calcareous shells of organisms. The organic carbon cycle (Figure 3.1-4:) only became effective with the onset of life about 3.5 billion years ago with photosynthesis and with the production of large biomass.

The two carbon cycles, inorganic and organic, are linked via CO_2 (Figure 3.1-2).

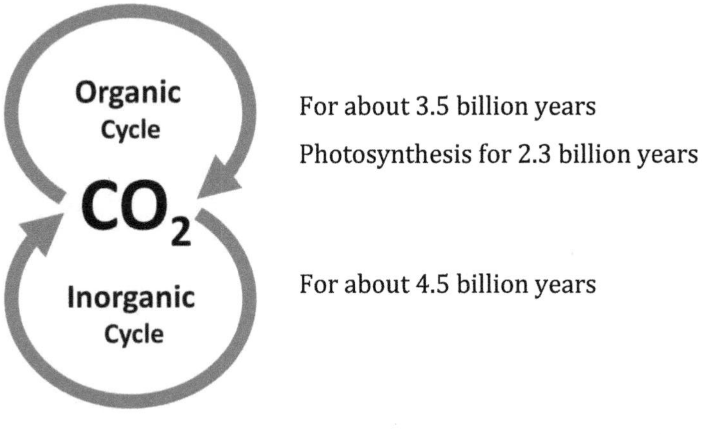

Figure 3.1-2: Linking the inorganic with the organic Carbon cycle via the CO_2

3.1.2 The inorganic carbon cycle

The inorganic (geological) carbon cycle begins with the formation of the earth about 4.5 billion years ago.

Inorganic Carbon is mainly found distributed as follows:

- gaseous as CO_2 in the atmosphere
- dissolved in water as CO_2, H_2CO_3, HCO_3^-, CO_3^{2-}
- in carbonate rocks and sediments
- as elemental carbon

Transformations between carbon forms occur through chemical reactions or through tectonic processes in the earth's interior.

The occurrences of inorganic carbon in the exogenous spheres are interconnected through the water cycle. In the earth's interior, the exchange processes again run via tectonic processes.

Regulation of the air's carbon dioxide content via the inorganic carbon cycle (Figure 3.1-3:) proceeds slowly over geologic time. The CO_2 in the air either dissolves directly in the surface water or reaches the soil and water bodies dissolved in the rainwater.

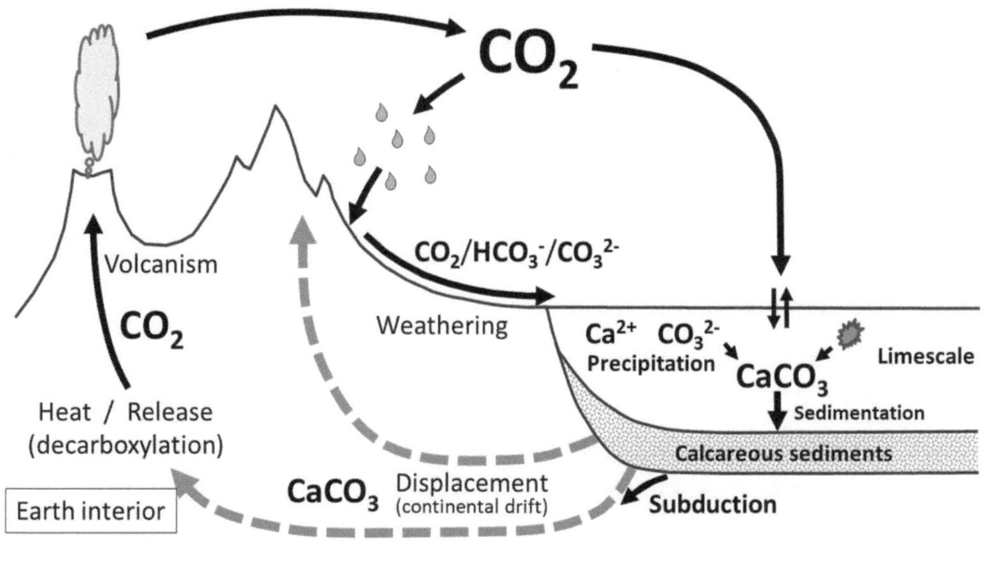

Figure 3.1-3: Inorganic carbon cycle

By dissolving in water, hydrogen carbonate ions (HCO_3^-) and to a lesser extent carbonate ions (CO_3^{2-}) as well as protons (H^+) are formed in addition to the gaseously dissolved CO_2 (chap. 4.1).

In water bodies, if the solubility product of lime ($CaCO_3$) (chap. 4.7.2) is exceeded, lime precipitation may occur, with subsequent sedimentation of the lime particles. Some aquatic organisms incorporate lime into their shells (e.g., foraminifera, calcareous algae). After the organisms die, these calcareous shells sediment. Both processes can and could form extensive calcareous sediments (especially in the primordial oceans).

Due to tectonic processes (continental drift) large amounts of calcareous sediments entered the earth's interior or were folded up as mountains (e.g. Limestone Alps, Dolomites). During the displacement of limestone into the earth's interior (subduction), which still occurs today at the edges of the continental plates, CO_2 can be split off at high temperatures (thermal decarboxylation). The carbon dioxide is then released back into the atmosphere via volcanism.

Precipitated lime and carbonate rock can re-dissolve through weathering processes and lead to the release of carbonate ions, hydrogen carbonate ions and CO_2.

Disturbances of the equilibrium in the inorganic carbon cycle can occur due to large volcanic eruptions (supervolcanoes). Up to 0.54 Gt CO_2-C (= 1.98 Gt CO_2) is released to the atmosphere per year via volcanism under normal conditions [25].

Anthropogenic processes increased the atmospheric CO_2 content by an average of 2.41 ppmv per year over the period 2010-2020, corresponding to an increase of 5.2 Gt carbon C (19.2 Gt carbon dioxide CO_2) per year (chap.3.1.5).

3.1.3 The organic carbon cycle

The organic carbon cycle has only existed since the beginning of life on earth about 3.5 billion years ago.

Organic Carbon is found distributed primarily as follows:

- Living and dead biomass
- Fossil deposits (lignite, hard coal, oil, natural gas (methane))
- Methane in the atmosphere
- nowadays: Plastic waste

In the organic carbon cycle (Figure 3.1-4:) inorganic CO_2 is the basis for the buildup (assimilation) of organic matter, which is generally summarized as biomass.

The organic biomass is biologically degraded again under oxygen consumption (aerobic) to CO_2. In the absence of oxygen (anaerobic), organic matter is reduced to methane CH_4 (see also Figure 5.7-8:), which can escape into the atmosphere where it acts as a greenhouse gas. In the presence of oxygen, it can be oxidized to CO_2.

Part of the biomass is deposited and transformed in geological periods (peat, lignite , hard coal, oil , natural gas).

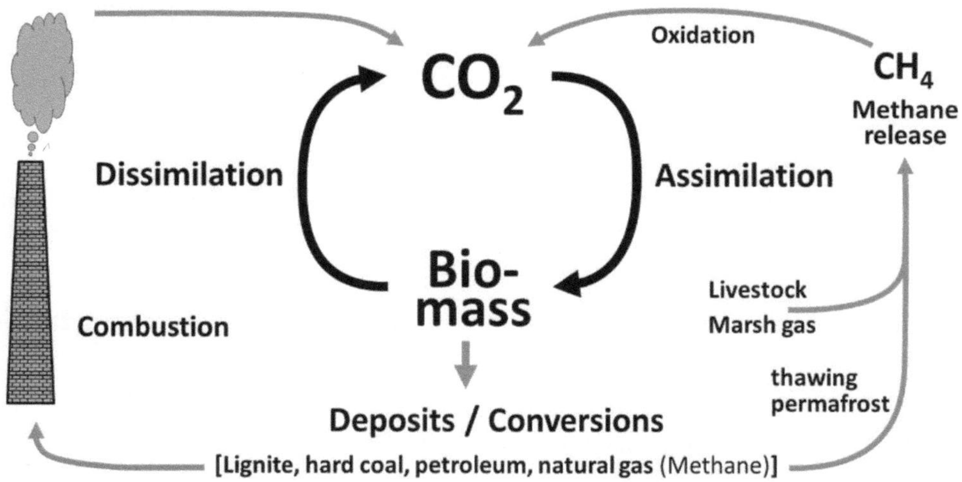

Figure 3.1-4: Organic carbon cycle

Organic biomass is reduced compared to CO_2 and is therefore rich in energy. This means that energy can be obtained through the oxidation of these compounds. This is done by all heterotrophic organisms that live on this energy. During respiration and decomposition of organic matter (dissimilation), inorganic CO_2 is released again.

The same thing happens when fossil fuels are burned. Through combustion (oxidation), energy is gained and CO_2 is released again.

The current burning of fossil fuels therefore releases a large proportion of the CO_2 deposited in the form of organic matter, thus disrupting the natural cycle. As a result, the concentration of CO_2 in the atmosphere, which has remained almost constant for thousands of years, increases.

The content of CO_2 in the atmosphere is influenced by:

Photosynthesis	=	Consumption of CO_2
Cellular respiration (living beings)	=	Release of CO_2
Fossil fuel combustion	=	Release of CO_2
Volcanism	=	Release of CO_2
Solution in water	=	Binding / release of CO_2

Figure 3.1-5: Influences on the CO_2 content of the atmosphere

The gaseous CO_2 may be assimilated and fixed in organic matter or precipitated as a salt of carbonic acid in the form of hydrogen carbonates or carbonate and fixed as sediment or rock.

The bound CO_2 can be released by decomposition processes of organic matter or by conversion of hydrogen carbonates or carbonates by acid action

The CO_2 concentration of the atmosphere is of central importance for the climate. Therefore, the conversion processes (Figure 3.1-5:) are particularly relevant here.

3.1.4 Measurements of the CO_2 content of the atmosphere

Until 800,000 years ago, data on the CO_2 content of the atmosphere are based largely on assumptions, models and geological findings. Measurements are available for the period from 800,000 years onwards from the study of trapped air bubbles in ice cores in eastern Antarctica. Since 1958, the CO_2 concentration has been measured directly in the atmosphere. These measurements were started on Mauna Loa in Hawaii [31].

Ice cores

The first project to obtain CO_2 data from an ice core was conducted at *Law Dome* in eastern Antarctica. Data for the past 1000 years were obtained there [38]. The measurements show the impact of the onset of industrialization in the early 19th century on atmospheric CO_2 levels. Three subprojects resulted in a total of measured values from the year 1006 to 1978 (Figure 3.1-6). It can be seen that the value of CO_2 fluctuated around 280 ppmv in the 800 years before 1800 (274-284 ppmv).

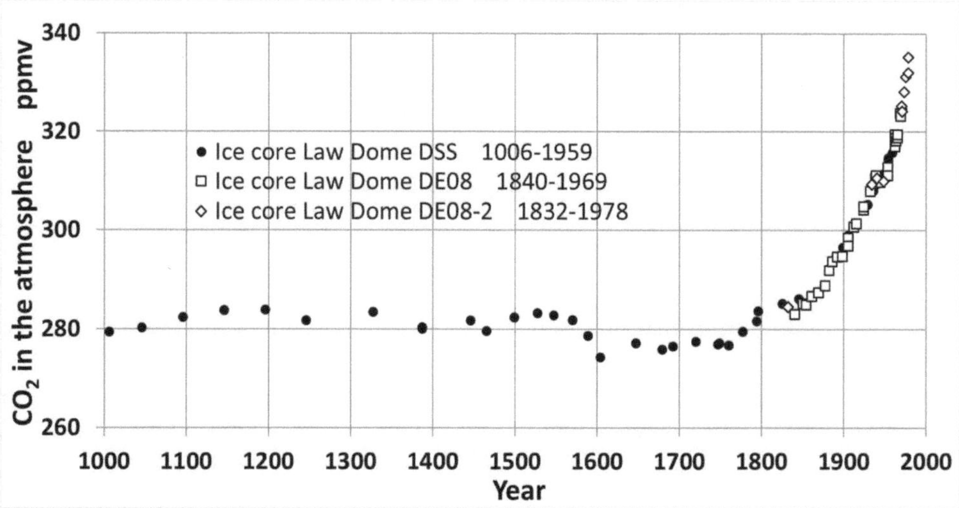

Figure 3.1-6: CO_2 readings from the Law Dome ice drilling project. Data: [38]

Between 1800 and 1850, the value began to increase with the onset of industrialization, and then exceeded 300 ppmv at the beginning of the 20th century (1912).

An initial increase in atmospheric CO_2 concentrations began as early as the mid-18th century due to the release of carbon dioxide from changes in land use. With the onset of industrialization in the mid-19th century, the burning of fossil fuels through the associated release of CO_2 greatly accelerated the rise of CO_2 concentration in the atmosphere.

The *Vostok Ice Core* well, which followed in time, yielded values for CO_2 in the atmosphere up to 420.000 years ago. This project also shows that prior to the onset of industrialization around 1800, no levels of CO_2 in the atmosphere above 300 ppmv were measured in the 420,000 years before that. During the 420,000 year measurement period, 4 ice ages and 4 interglacial periods were recorded with a range of variation in CO_2 values from 180 to 280-300 ppmv [39].

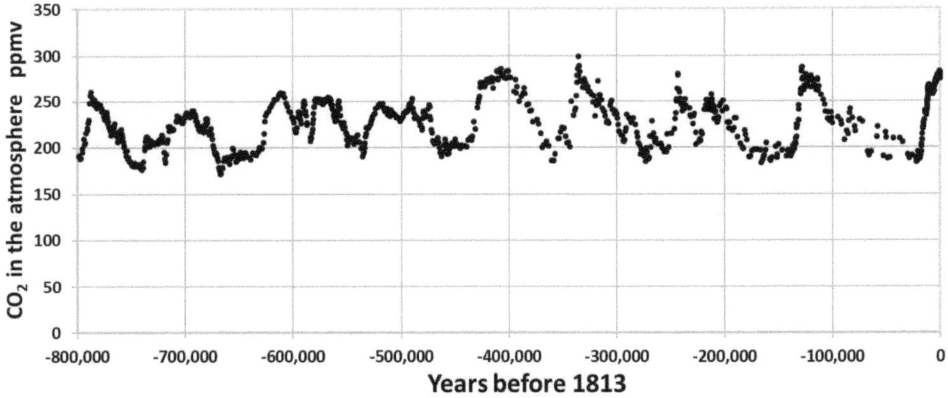

Figure 3.1-7: Atmospheric CO_2 levels over 800,000 years, EPICA project Dome C Data: [40]

The following and currently furthest back measured values for the CO_2 content of the atmosphere were obtained from the *EPICA project* in *Dome C*, Antarctica [40]. From these measurements, it appears that over a period of 800,000 years prior to the onset of industrialization, the CO_2 content of the atmosphere was always below 300 ppmv (max. 298.6 ppmv) (Figure 3.1-7). During this period there were 8 ice ages (minimum) and 9 warm ages (maximum).

Another ice core drilling, which will provide information about the last 1.2 million years, is planned.

In summary, during the last 800,000 years before 1912, the CO_2 content of the atmosphere was always below 300 ppmv, predominantly in the range of 280 ppmv. In the following period, it increased to 418 ppmv (2022) in 110 years.

3.1.5 Current situation

In the pre-industrial period (until about 1750), the CO_2 content in the air was still within a normal range of variation between 275 and 284 ppmv [41], averaging about 280 ppmv, as determined by the study of ice cores (chap. 3.1.4).

Direct measurements have been made in the atmosphere since 1958. The absorption of IR radiation by the CO_2 molecule in the gas phase is used as the measurement method. First measurements started on Mauna Loa, Hawaii, so the longest series of measurements for CO_2 is also available from there.

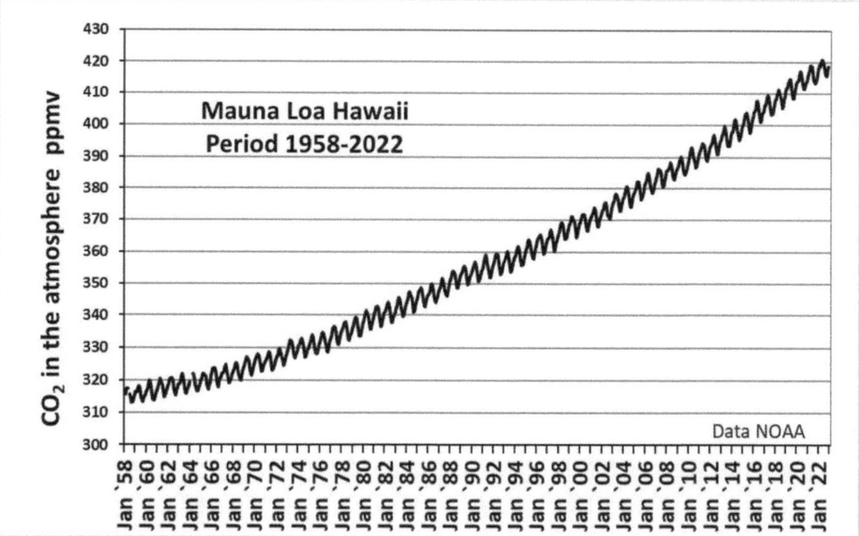

Figure 3.1-8: CO_2 concentration in the atmosphere in ppmv on Mauna Loa, Hawaii. Data NOAA [31]

In Figure 3.1-8:, monthly mean CO_2 values for the period 1958 to 2022 are shown from Mauna Loa, Hawaii (Data NOAA [31]). The curve shows a steady increase in concentration superimposed by seasonal fluctuations. In the spring of 2022, 420 ppmv was exceeded for the first time. The mean value for 2022 was 418.56 ppmv. Compared to the pre-industrial period before 1750 at 280 ppmv, this represents a 50% increase.

Linking these data obtained by direct measurements of CO_2 in the atmosphere from 1958 onward with the Law Dome project going back to the year 1000, one can see the sudden and sharp increase of CO_2 in the atmosphere from about the year 1800 (Figure 3.1-9:). It also shows the good overlay of data from the different projects.

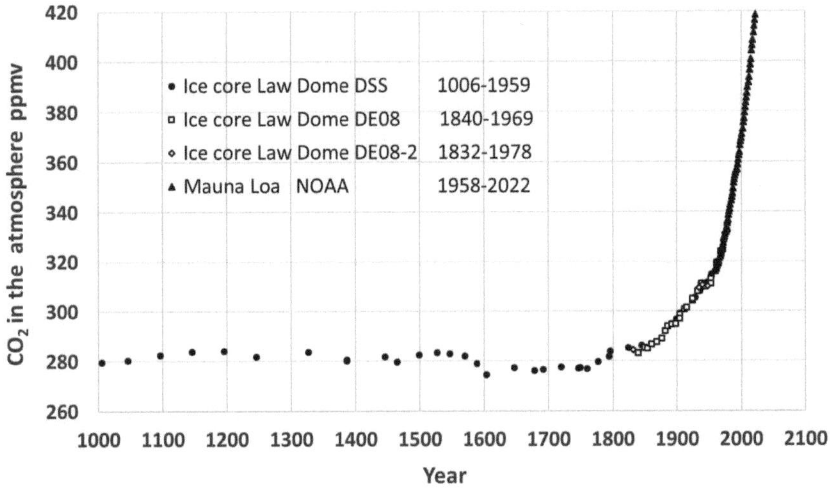

Figure 3.1-9: CO_2 in the atmosphere since the year 1000
Ice Core Project Law Dome and Mauna Loa (NOAA)

Looking at the increase in annual mean CO_2 concentration in 10-year increments from 1970 to 2020 (Figure 3.1-10:), each averaged over 10 years, one can see an acceleration of the increase from 1.31 ppmv/a in 1970/1980 to 2.41 ppmv/a for the period 2010/2020.

The mean increase of 2.41 ppmv/a from 2010-2020 corresponds to an increase of 5.2 Gt of carbon C or 19.2 Gt of carbon dioxide CO_2 per year in the atmosphere.

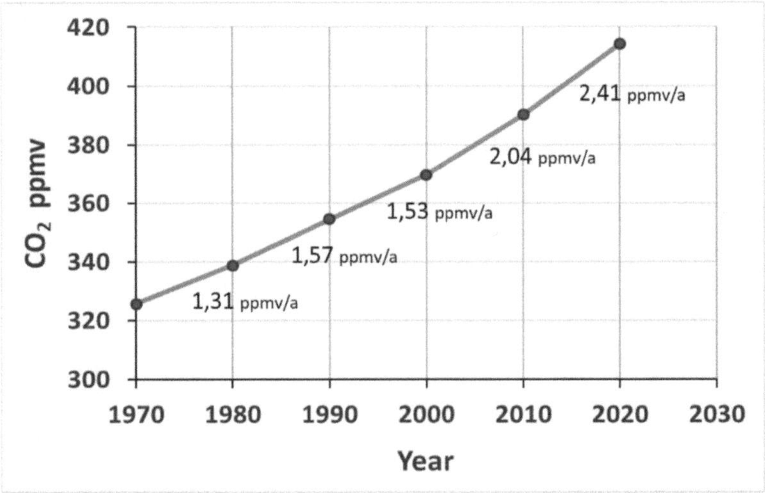

Figure 3.1-10: Mean CO_2 concentration in 10-year steps from 1970 to 2020

For Germany, measurements on the Schauinsland (Black Forest) show a development comparable to Mauna Loa for the annual mean values of the CO_2 concentration in the atmosphere (Table 3.1-2).

Table 3.1-2: Annual mean values for the carbon dioxide content of the air in vol% at Schauinsland [42] and on Mauna Loa [29]

	Schauinsland	Mauna Loa
1972	0.0330	0.0327
1982	0.0343	0.0341
1992	0.0357	0.0356
2002	0.0375	0.0373
2012	0.0395	0.0394

1 ppmv (mL/m^3) = 0.0001 vol% or 1 vol% = 10,000 ppmv

The example calculations in this paper are predominantly based on a CO_2 content of 0.0420 vol% (420 ppmv) in the atmosphere reached in 2022.

The calculations for the dissolution of the gas in water are usually carried out for a water temperature of 10 °C as an example, since this corresponds approximately to the annual mean temperature in Germany and thus also to the temperature of the groundwater in the area of the isothermal zone (approx. 15 m deep).
The values for the CO_2 molar concentration $c(CO_2)$ and the mass concentration $\beta(CO_2)$ in the atmosphere at a content of 420 ppmv (reached in April 2022) at normal pressure and 10 °C are given in Table 3.1-3 (calculation example chapter 10.2).

Table 3.1-3: CO_2 concentration in the air CO_2 (10 °C) at a measured value of 420 ppmv

$c(CO_2)$ =	$1.81 \cdot 10^{-5}$	mol/L (air)
$\beta(CO_2)$ =	0.796	mg/L (air)

This increasing concentration in the air over the years directly affects the amount of CO_2 dissolved in surface waters. Since about 70% of the earth's surface consists of water, these exchange processes play a major role. Also affected is the pH of the water bodies, which decreases due to the increased CO_2 input (section 4.2). In the world's oceans, this particularly affects organisms with calcareous skeletons, e.g. corals.

3.2 Carbon dioxide in the soil air

Air is found in the gaps and pore spaces of the soil and rock, isolated from atmospheric air in deeper layers. This soil air is in contact with percolating water and groundwater. The CO_2 content of soil air is usually higher than in atmospheric air.

The top layer of soil is usually enriched with humus substances. Here, intensive biodegradation of organic matter takes place, releasing CO_2. As the gas exchange between soil air and atmosphere decreases with increasing depth, CO_2 accumulates in the soil air.

Infiltrating Rainwater infiltrates vertically into the subsurface through the gap spaces in the soil. This water was originally in contact with atmospheric air, so that CO_2 could dissolve in the water accordingly. At an atmospheric concentration of 420 ppmv (spring 2022), this would be 0.994 mg CO_2 per liter in solution equilibrium at 10°C (4.1).

Due to the higher concentration and the resulting higher partial pressure of CO_2 in the soil air, more CO_2 dissolves in the leachate . The gaseously dissolved CO_2 is displaced into the depths with the seepage water. The soil air below the humic zone is in gas exchange with the seepage water, so that the CO_2 content of the soil air also increases in deeper layers.

The dissolution of CO_2 in the leachate decreases the pH (chap. 4.2 and 10.5) and increases the total carbon content (the carbonic acid sum) of the water. However, the CO_2 share may decrease due to the dissolution of lime and the associated increase in pH, which leads to a shift in the carbonic acid forms (chap. 4.3).

The leachate finally passes into the groundwater body, which accumulates over impermeable layers and shows a horizontal flow movement. Here, too, gas exchange occurs with the soil air located above the groundwater body.

Spring water is emergent groundwater and therefore usually has an increased CO_2 content compared to the solution equilibrium with atmospheric air. Therefore, CO_2 is outgassed from the water in the source area and a pH increase occurs in the water (chap. 5.5).

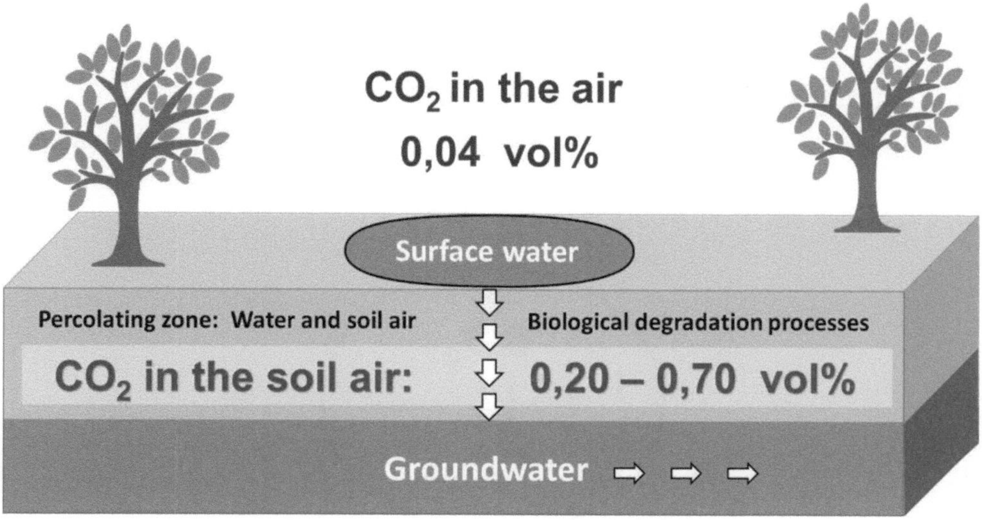

Figure 3.2-1: Carbon dioxide in the soil air

The absolute content of CO_2 in the soil air depends not only on the biological activity but also strongly on the porosity of the soil and the possibility of gas exchange with the atmosphere. One usually finds values of 0.2 - 0.7% CO_2 by volume in the soil air. Coarse-pored soils generally have lower CO_2 concentrations than fine-pored soils. Clay-rich, poorly drained soils can have values up to over 5%. [43]

4 The carbonic acid system in water

4.1 Solution of gaseous CO_2 in water

Dissolving CO_2 in water is a physical process. This is a gas exchange between the air and the water. CO_2 from the air passes into the water and at the same time CO_2 molecules also pass from the water into the air.

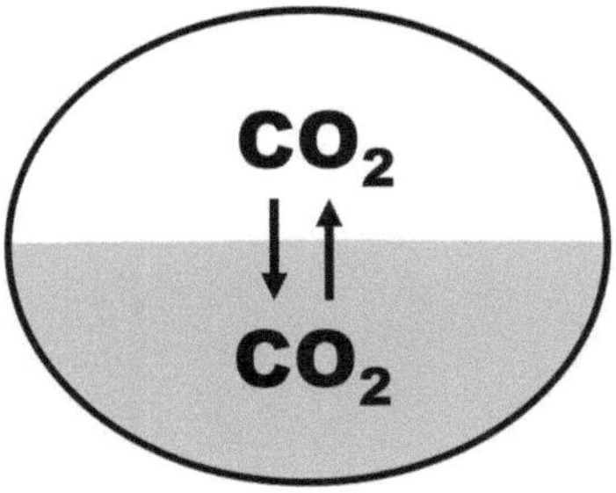

Figure 4.1-1: Gas exchange of CO_2 between air and water

An equilibrium is reached between the two processes when, per unit of time, just as many CO_2 molecules migrate from the air into the water as from the water into the air in the opposite direction. In the air and in the water, the corresponding equilibrium concentrations are thus established, which correspond to the existing physical conditions (e.g. temperature, pressure).

This process can be represented as an equilibrium reaction.

$$CO_{2(g)} \rightleftharpoons CO_{2(w)} \qquad \text{(eq. 4.1-1)}$$

(g) = in the gas phase
(w) = dissolved in water

The general laws for the dissolution of gases in water apply. The calculations are performed as a function of partial pressure, temperature and, if necessary, ionic strength of the water (chap. 0).

$CO_{2(w)}$ dissolved in water reacts chemically with water to form carbonic acid H_2CO_3 (chap. 4.3). However, carbonic acid is only present in water in very low concentrations and cannot be clearly detected analytically (chap. 4.4). Therefore, carbonic acid is combined with gaseous dissolved carbon dioxide to form the expanded carbon dioxide and used as the basis for calculations in the carbonic acid system (Section 4.4).

Therefore, in water chemistry, different content specifications for the carbon dioxide in water are distinguished:

Carbon dioxide dissolved in water in gaseous form: $CO_{2(w)}$

The expanded carbon dioxide: CO_2 $(= CO_{2(w)} + H_2CO_3)$

(The expanded CO_2 is used in this paper ***without an index.*** *It is often referred to as $CO_{2(aq)}$ in other publications)*

The constants for calculating the dissolved CO_2 in water always apply to the expanded CO_2, i.e. including carbonic acid. Also, when calculating the forms of carbonic acid using the mass action constants, the extended carbon dioxide is usually always calculated. To specify the purely gaseous dissolved $CO_{2(w)}$, one would have to calculate this value separately using the corresponding mass action constants.

In **fresh water** solubility calculations, the influence of salinity can be neglected in practice (approximation: Ionic strength I = 0).

With **seawater** you have to take into account the high salt content, which leads to a lower solubility of CO_2. A chlorinity of 19‰ was used for the calculations of salt water [44], which corresponds to the mean salinity of the oceans of 35‰.

The calculated values listed in Table 4.1-1 are equilibrium concentrations in water which occur in solution equilibrium with the gas phase. For example, this applies to freshwater for rainwater or surface water in exchange with atmospheric air. (Example calculation 0)

Table 4.1-1: CO_2 mass concentration in water at solution equilibrium with atmospheric air (10°C), 420 ppmv CO_2

Temperature	**CO_2 in water in mg/L**	
°C	Freshwater I = 0	Seawater 19‰ Cl^-
0	1.44	1.17
5	1.19	0.974
10	**0.994**	0.823
15	0.843	0.705
20	0.724	0.610
25	0.629	0.534
30	0.536	0.472

As shown above, the values given in the tables apply to the expanded CO_2, i.e. to the sum $CO_{2(w)}$ + H_2CO_3. The deviation from the value of the purely gaseous dissolved $CO_{2(w)}$ is about 0.2 %.

At a fraction of 420 ppmv CO_2 in the atmosphere, 0.994 mg/L CO_2 dissolved in water is present in solution equilibrium at normal pressure of 1013.25 hPa (1.01325 bar) and 10 °C (Table 4.1-1) (calculation example chap. 0).

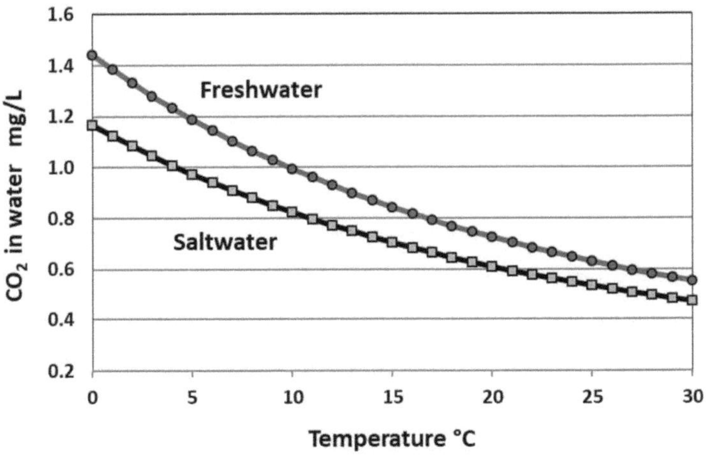

Figure 4.1-2: CO_2 in freshwater and in saltwater in solution equilibrium in mg/L (CO_2 atmosphere 420 ppmv, 10 °C)

An increase in temperature in the water results in lower solubility (lower equilibrium concentration) of CO_2 in the water. It can be seen that the solubility of CO_2 (equilibrium concentration) in water decreases by about half from 0 to 20 °C (Table 4.1-1, Figure 4.1-2).

If there are higher concentrations and thus higher partial pressures of CO_2 in the gas phase (Table 4.1-2), for example due to biological degradation processes in the soil or in deeper water layers of standing waters, more CO_2 can dissolve in the water depending on the higher partial pressure. Soil air usually contains 0.2 - 0.7% CO_2 by volume [45] in clay-rich, poorly drained soils, however, it can exceed 5% (calculation example, chap. 10.4). Groundwater formed in contact with soil air therefore has elevated concentrations of CO_2 compared to solution equilibrium with the atmosphere (Table 4.1-2). Therefore, sources release CO_2 accordingly until the solution equilibrium with the air is restored.

Similarly, deep water from lakes generally contains higher levels of CO_2 due to biological degradation processes at depth (chap. 5.7).

Table 4.1-2: CO_2 contents in water as a function of the partial pressure of CO_2 in the gas phase, calculated for 10 °C.

$CO_{2(g)}$ vol%	Partial pressure bar	CO_2 in water in mg/L (I=0)
0.2	0.00203	4.7
0.7	0.00709	16.6
1	0.0101	23.7
3	0.0304	71.0
5	0.0507	118.3

1 ppmv = 0.0001 vol% or 1 vol% = 10,000 ppmv

4.2 CO_2 and the pH value of natural waters

Annotation: These calculations for the pH value only apply to unbuffered water. Buffer systems attenuate pH changes.

Depending on the amount of CO_2 dissolved in the water, the chemical reaction of the CO_2 with the water changes the pH value. When CO_2 reacts with water, protons are released (chap. 4.3) and the pH value decreases accordingly.

Natural rainwater, which has become saturated with CO_2, has a pH value of 5.556 at an atmospheric content (2022) of 420 ppmv CO_2 (calcul. example chap. 10.5).

Table 4.2-1: Calculated pH values of rainwater from 1796 to 2020
CO_2 data: 1796 to 1944 Law Dome [38], then NOAA [31]

Year	CO_2 ppmv	pH
1796	284	5.641
1825	285	5.640
1850	285	5.640
1877	289	5.637
1899	297	5.631
1924	304	5.626
1944	310	5.622
1960	317	5.617
1980	339	5.602
2000	370	5.583
2020	414	5.559

This calculable pH of rainwater caused by CO_2 has decreased from 5.64 to 5.56 since the beginning of industrialization (1800) until today due to the enrichment of CO_2 in the atmosphere (Table 4.2-1) (calculations for 10°C).

The actual pH value of rainwater can be significantly lower if other acidifying gases such as SO_2 or NO_2 are dissolved in the rainwater!

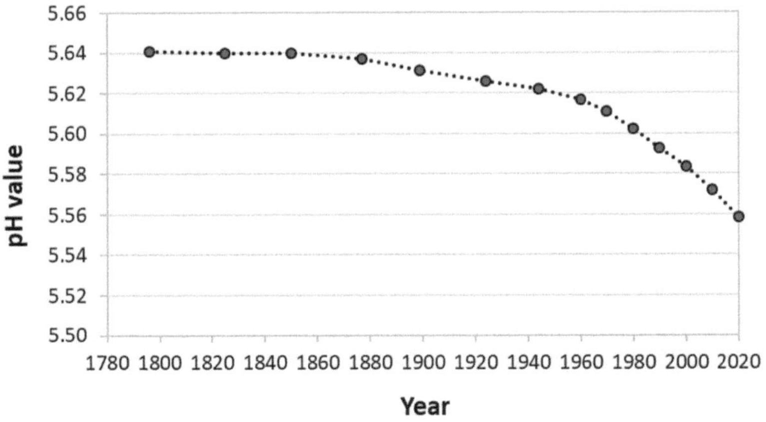

Figure 4.2-1: Calculated theoretical pH values of rainwater from 1796 to 2020. CO_2 data: 1796 to 1944 Law Dome [38], then NOAA [31]

In seawater the increase in CO_2 concentration in the atmosphere also leads to a decrease in pH. However, the extent of pH reduction can be mitigated compared to freshwater by the presence of buffer systems (e.g., borates).
In the soil, higher partial pressures lead to a stronger acidification of the leachate (Table 4.2-2).

Table 4.2-2: Calculated theoretical pH values of water in contact with soil air (For comparison, the value for the atmosphere)

	vol% CO_2	pH
Atmosphere	0.042	5.56
Soil air	0.2	5.22
	0.7	4.94
	1	4.87
	2	4.72
	3	4.63
	5	4.52

1 ppm = 0.0001 vol% or 1 vol% = 10,000 ppm

However, the pH values calculated as a function of the composition of the soil air are only theoretical and represent the acidification potential.

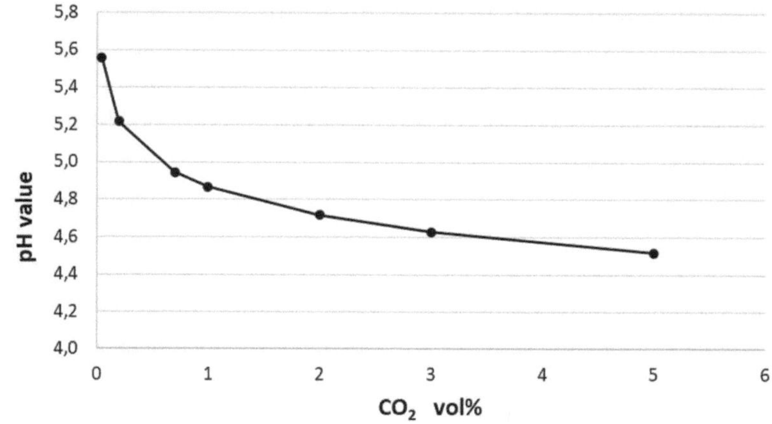

Figure 4.2-2: Calculated theoretical values for the pH value as a function of the CO_2 content of the soil air

Groundwaters generally have pH values above 6. Thus, processes are present that at least partially compensate for the pH reduction in the soil. These are, for example, chemical reactions, ion exchange processes or the dissolution of lime. (see also chap. 5.4)

4.3 Chemical reaction with water - the carbonic acid forms

The CO_2 physically dissolved in the water in gaseous form reacts chemically with the water, forming carbonic acid (H_2CO_3):

$$CO_{2(w)} + H_2O \rightleftharpoons H_2CO_3 \quad \text{(hydration)} \qquad \text{(Eq 4.3-1)}$$

($CO_{2(w)}$ = anhydride of carbonic acid)

Carbonic acid is a biprotic weak acid with 2 dissociation stages. It dissociates in the first step to hydrogen carbonate and releases a proton (H^+):

$$H_2CO_3 \rightleftharpoons HCO_3^- + H^+ \qquad \text{(Eq 4.3-2)}$$

The hydrogen carbonate ion (HCO_3^-) formed during the dissociation of carbonic acid is also a weak acid and can dissociate further at higher pH values (2. Dissociation) and forms carbonate ions (CO_3^{2-}) and H^+:

$$HCO_3^- \rightleftharpoons CO_3^{2-} + H^+ \qquad \text{(Eq 4.3-3)}$$

In an **open system**, however, these chemical dissociation reactions have no effect on the equilibrium concentration of gaseously dissolved CO_2 as long as re-equilibration in contact with air is possible. If gaseously dissolved CO_2 is lost through chemical reaction, CO_2 from the air continues to dissolve in the water until the original equilibrium state for the gaseously dissolved CO_2 is restored. In equilibrium, the concentration of gaseously dissolved CO_2 depends only on the solution conditions (temperature, partial pressure) and is not influenced by the chemical reaction.

$$\begin{array}{l} \text{Air} \\ \downarrow\uparrow \\ CO_{2(w)} + H_2O \rightleftharpoons H_2CO_3 \rightleftharpoons HCO_3^- + H^+ \rightleftharpoons CO_3^{2-} + 2H^+ \end{array} \qquad \text{(Eq 4.3-4)}$$

CO_2, carbonic acid, as well as hydrogen carbonate and carbonate are simplified in the sum as *carbonic acid* and accordingly as carbonic acid forms. Thus, the forms of carbonic acid shown in Table 4.3-1 can coexist in water.

Table 4.3-1: The forms of carbonic acid in water

Carbonic acid - forms:	
$CO_{2(w)}$	Carbon dioxide dissolved as gas
H_2CO_3	Carbonic acid
HCO_3^-	Hydrogen carbonate ion
CO_3^{2-}	Carbonate ion

A molar balance can be established for the sum of the carbonic acid forms (molar concentrations c(i)). The molar balance presented here is referred to as $c(C)_{real}$ to distinguish it from the mass balance in the simplified system in chapter 4.4.

<u>Molar balance:</u>

$$c(C)_{real} = c(CO_{2(w)}) + c(H_2CO_3) + c(HCO_3^-) + c(CO_3^{2-}) \qquad \text{(Eq 4.3-5)}$$

4.4 The simplified carbonic acid system

The carbonic acid and the carbonate ion are present in the pH neutral range only in very low concentrations and are not directly detectable analytically, or only with difficulty.

Knowledge of the concentration of the carbonate ion cannot be dispensed with because of the solubility product of $CaCO_3$. One therefore calculates the concentration from the other components of the system (see below).

However, in the case of carbonic acid itself, we can simplify because its concentration is not needed for the calculations in the carbonic acid system. Since only about 0.2% (see also section 1.4) of it is formed from the $CO_{2(w)}$ physically dissolved in water, both compounds are simply represented in the sum as expanded CO_2 [6] and they are also recorded analytically together (see also section 4.1):

$$CO_2 = CO_{2(w)} + H_2CO_3 \qquad \text{(Eq 4.4-1)}$$

As a consequence of the simplification, one also summarizes the reaction equations (Eq. 4.3-1) and (Eq. 4.3-2) and represents it as if HCO_3^- is formed directly from CO_2 and H_2O, without the intermediate H_2CO_3.

(Eq 4.3-1) + (Eq. 4.3-2) : $$CO_2 + H_2O \rightleftharpoons HCO_3^- + H^+ \qquad \text{(Eq 4.4-2)}$$

Thus, taking this simplification into account, the carbonic acid system in water looks as shown in Figure 4.4-1:.

Simplified Carbonic Acid System

$$CO_2 + H_2O \rightleftharpoons HCO_3^- + H^+$$

$$HCO_3^- \rightleftharpoons CO_3^{2-} + H^+$$

Figure 4.4-1: Simplified carbonic acid system

The overall equation is then in the simplified carbonic acid system:

Air

$\downarrow\uparrow$

$$CO_2 + H_2O \rightleftharpoons HCO_3^- + H^+ \rightleftharpoons CO_3^{2-} + 2\,H^+ \qquad \text{(Eq 4.4-3)}$$

For the carbonic acid forms, the carbonic acid itself is omitted in the simplified system (Table 4.4-1:).

Table 4.4-1: The carbonic acid forms in the simplified system

Carbonic acid shapes in the simplified system:	
CO_2	expanded carbon dioxide ($CO_{2(w)}$ + H_2CO_3)
HCO_3^-	Hydrogen carbonate ion
CO_3^{2-}	Carbonate ion

The sum equation (molar balance) of inorganic carbon in water is c(C) with the simplification described above:

Carbonic acid sum (Molar balance):

$$c(C) = c(CO_2) + c(HCO_3^-) + c(CO_3^{2-}) \quad \text{(Eq 4.4-4)}$$

This carbonic acid sum (= total concentration of inorganic carbon C in water) is marked with different signs in the literature.

Formula symbol:	CT = CTotal	[46] [44]
	QC = Quantum-C	[47] [48]
	[C]	[49]
	c(C)	[50] [51]
	c(DIC)	[52]

In the rest of this paper, the carbonic acid sum is always represented as c(C).

As an analytical value, the carbonic acid sum is given either as DIC or TIC (chap. 7.2.1).

Analytical values:

DIC = Dissolved Inorganic Carbon [52]

TIC = Total Inorganic Carbon [In this paper.]

4.5 The Equilibrium constants

4.5.1 Formation of carbonic acid (hydration)

Considering the rate in the hydration reaction (Eq. 4.5-1), it is found that the formation of H_2CO_3 (hydration) is much slower than the reverse reaction, the splitting of HCO_3^- into $CO_{2(w)}$ and H_2O (dehydration).

$$CO_{2(w)} + H_2O \underset{k_2 = 20 sec^{-1}}{\overset{k_1 = 0.03 sec^{-1}}{\rightleftarrows}} H_2CO_3 \qquad (25\ °C) \qquad \text{(Eq 4.5-1)}$$

(k_1 ;k_2 = rate constants of the reaction)

The reverse reaction (k_2) proceeds 666 times faster than hydration (k_1) [53].

For the reaction equilibrium, this means that the equilibrium is strongly on the left side of the reaction, i.e., there is a lot of $CO_{2(w)}$ in the water and very little H_2CO_3.

The mass action equation for this reaction is:

$$\frac{a(H_2CO_3)}{a(CO_{2(w)})} = K = 2 \cdot 10^{-3} \rightarrow a(H_2CO_3) = 0.002 \cdot a(CO_{2(w)}) \quad [1] \quad \text{(Eq 4.5-2)}$$

(the activity of the water is considered constant and is included in the constant See: Basics Mass Action Law, chap. 11.3)

Or to put it another way:

Ratio $CO_{2(w)}$: H_2CO_3 like 500: 1 or 0.2% H_2CO_3

The low reaction rate is noticeable, for example, analytically in the determination of the base capacity (section 7.2.2.2), since one must always wait for the reaction equilibrium to be re-established (a few seconds) after OH^- addition. All other reactions occur spontaneously, without noticeable delay.

4.5.2 The dissociation constants of carbonic acid

From the real (true) dissociation of carbonic acid

$$H_2CO_3 \rightleftharpoons HCO_3^- + H^+ \qquad \text{(Eq 4.5-3)}$$

derives the real (true) 1. Dissociation constant K_1' off.

$$\frac{a(HCO_3^-) \cdot a(H^+)}{a(H_2CO_3)} = K_1' \qquad \text{(Eq 4.5-4)}$$

This dissociation constant is not needed for normal calculations in the carbonic acid system.

The constants of the simplified system are used as the basis for calculations.

1. Dissociation constant K_1

$$CO_2 + H_2O^* \rightleftharpoons HCO_3^- + H^+ \qquad \text{(Eq 4.5-5)}$$

$$\frac{a(HCO_3^-) \cdot a(H^+)}{a(CO_2)} = K_1 \qquad \text{(Eq 4.5-6)}$$

2. Dissociation constant K_2

$$HCO_3^- \rightleftharpoons CO_3^{2-} + H^+ \qquad \text{(Eq 4.5-7)}$$

$$\frac{a(CO_3^{2-}) \cdot a(H^+)}{a(HCO_3^-)} = K_2 \qquad \text{(Eq 4.5-8)}$$

*) The activity of the water $a(H_2O)$, which should actually be in the denominator of the fraction, is considered constant and included in the constant as usual ($K_1 = K \cdot a(H_2O)$). (See also Fundamentals of Mass Action Law chap. 11.3)

Annotation: Since the constant K_1 is derived from the simplified system, it is sometimes also called apparent 1st Dissociation constant.

The equilibrium constants K_1 and K_2 are derived from the **activities** of the reactants and depend only on temperature and pressure {K(T,p)}. If we use the molar concentrations of the substances instead of the activities, we obtain the constants K_1^* and K_2^*, which additionally depend on the ionic strength {K*(T,p,I)}.

$$\frac{c(HCO_3^-) \cdot c(H^+)}{c(CO_2)} = K_1^* \qquad \text{(Eq 4.5-9)}$$

$$\frac{c(CO_3^{2-}) \cdot c(H^+)}{c(HCO_3^-)} = K_2^* \qquad \text{(Eq 4.5-10)}$$

With the activity coefficient f_K of the respective constant, the formulaic relationship between K and K* results (Fundamentals of the Mass Action Law Chap. 11.3).

$$K^* = K \cdot f_K \qquad \text{(Eq 4.5-11)}$$

The constants K_1 und K_2 can be calculated as a function of temperature (Chap. 7.1)(Table 4.5-1).

Table 4.5-1: Dissociation constants K_1 and K_2 as a function of temperature (see also table 12.3 in the appendix)

Temperature in °C	K_1 in mol/L	K_2 in mol/L
0	$2.677 \cdot 10^{-7}$	$2.352 \cdot 10^{-11}$
5	$3.055 \cdot 10^{-7}$	$2.788 \cdot 10^{-11}$
10	$3.426 \cdot 10^{-7}$	$3.249 \cdot 10^{-11}$
15	$3.781 \cdot 10^{-7}$	$3.726 \cdot 10^{-11}$
20	$4.110 \cdot 10^{-7}$	$4.209 \cdot 10^{-11}$
25	$4.406 \cdot 10^{-7}$	$4.688 \cdot 10^{-11}$

In the system of equations of the dissociation constants are then the constants K_1 and K_2 are known by calculation.

$$\frac{a(H^+) \cdot a(HCO_3^-)}{a(CO_2)} = K_1 \qquad \text{(Eq 4.5-12)}$$

$$\frac{a(CO_3^{2-}) \cdot a(H^+)}{a(HCO_3^-)} = K_2 \qquad \text{(Eq 4.5-13)}$$

There remain **4 unknown quantities** in the two equations:

$a(H^+)$; $a(CO_2)$; $a(HCO_3^-)$; $a(CO_3^{2-})$

If two of the remaining quantities $a(CO_2)$, $a(HCO_3^-)$, $a(CO_3^{2-})$, $a(H^+)$ are known, then the other two can be calculated using the equations (2 equations, 4 unknowns):

➔ Two measurands are required to describe the system (in addition to temperature and ionic strength).

If one knows at least 3 measurands by analytics, one can check the analysis computationally, since each measurand can be calculated by the others and thus checked.

Because the generally very low concentration of carbonate ions in water is difficult or impossible to measure analytically, 3 of the 4 unknown quantities remain for practical use. In addition, however, it is possible to record the carbonic acid sum c(C), so that a total of 4 measured variables are available again, which can be used in different combinations (Table 4.5-2).

4 measurands: $a(H^+)$; $c(CO_2)$; $c(HCO_3^-)$; c(C)

Annotation: Analytically, the substance amount concentrations c are always recorded. Only pH measurement directly records the activity a.

Table 4.5-2: Possible combinations of the measured variables

Possible combinations		
c(C) / $a(H^+)$	⟶	TIC / pH
$c(HCO_3^-)$ / $a(H^+)$	⟶	$K_{S4.3}$ / pH
$c(CO_2)$ / $a(H^+)$	⟶	$K_{B8.2}$ / pH
$c(HCO_3^-)$ / $c(CO_2)$	⟶	$K_{S4.3}$ / $K_{B8.2}$

(For the designation of the measured variables, see chapter 7.2)

The most accurate calculations in the carbonic acid system are possible on the basis of TIC and pH, since the inaccuracies and systematic errors in the determination of acid and base capacity (chap. 7.2.2) are avoided.

4.5.3 Proportions of carbonic acid forms

When considering the proportions of carbonic acid forms, it is important whether the system is an **open** system with gas exchange or whether it is **closed to** the outside. The conditions in an open system are discussed in the thought model for carbonic acid in chap. 6.1.

If one considers the molar balance in a **closed** system without the possibility of gas exchange and without precipitation reactions,

Molar balance: $c(C) = c(CO_2) + c(HCO_3^-) + c(CO_3^{2-})$ (Eq 4.5-14)

so even if the **pH** in the system **changes**, the **carbonic acid sum** remains unchanged!

$$c(C) = \textbf{constant}$$

However, the **proportions** of the individual forms of carbonic acid change, corresponding to the equilibrium reactions.

$$CO_2 + H_2O \rightleftharpoons HCO_3^- + H^+ \rightleftharpoons CO_3^{2-} + 2\,H^+$$

Figure 4.5-1: Carbonic acid system closed off from gas exchange

Knowing the dissociation constants, it is possible to calculate the proportions $x(CO_2)$, $x(HCO_3^-)$ and $x(CO_3^{2-})$ of the carbonic acid sum in which the various forms of carbonic acid are present in the water as a function of pH value (calculation example 10.7).

$$x(CO_2) + x(HCO_3^-) + x(CO_3^{2-}) = 1 \qquad \text{(Eq 4.5-15)}$$

The proportions add up to 100%, or decimal = 1 (Table 4.5-3:).

Table 4.5-3: Relative proportions x(i) of the carbonic acid forms as a function of pH (10 °C), (I = 0) (see also Tab. 12.3 attached)

pH	$x(CO_2)$	$x(HCO_3^-)$	$x(CO_3^{2-})$
5.0	0.967	0.0331	0.000000108
6.0	0.745	0.255	0.00000829
7.0	**0.226**	**0.774**	**0.000251**
8.0	0.0283	0.969	0.00315
9.0	0.00282	0.966	0.0314
10.0	0.000220	0.755	0.245

As the pH increases, the proportion of CO_2 decreases, the proportion of HCO_3^- passes through a maximum, and the proportion of CO_3^{2-} increases.

This results in the **relative** distribution of carbonic acid forms for 10 °C shown in a linear plot in Figure 4.5-2:.

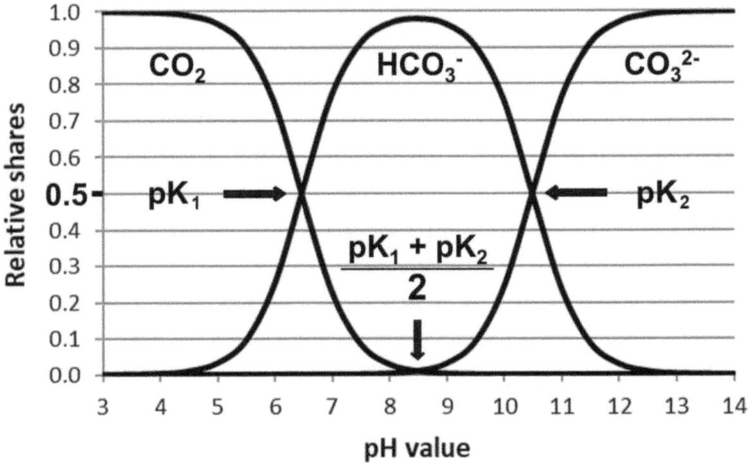

Figure 4.5-2: Linear representation of the carbonic acid propotions as a function of pH (10 °C) (I=0)

The corresponding pK values of the reactions are also shown in Figure 4.5-2:.

> Annotation:
> The negative decadic logarithm of the equilibrium constants K_1 and K_2 is denoted - as usual - as pK_1 and pK_2. The pK values are a measure of the strength of an acid (Basics 11.5). The relationship between pH and pK is given by the HENDERSON-HASSELBALCH equation (Eq. 4.5-16, derivation see chap. 11.6).
>
> $$pH = pK + \lg\frac{a(A^-)}{a(AH)} \qquad \text{(Eq 4.5-16)}$$

The linear representation clearly shows the importance of the intersections (Figure 4.5-2:). At pH = pK_1 = 6.47 (10 °C), 50% CO_2 and 50% HCO_3^- are present.

At pH = pK_2 = 10.49 (10 °C), HCO_3^- and CO_3^{2-} are then present at 50% each. The maximum of HCO_3^- (about 98%) occurs at pH = $(pK_1 + pK_2)/2$ = 8.48 with very low amounts of CO_2 and CO_3^{2-} (about 1% each). The representation is valid for 10 °C. In representations for 25 °C, which are often found in the literature, there are shifts to somewhat lower pH values.

More common, however, is the semi-logarithmic plot of the distribution of carbonic acid forms, as shown in Figure 4.5-3:.

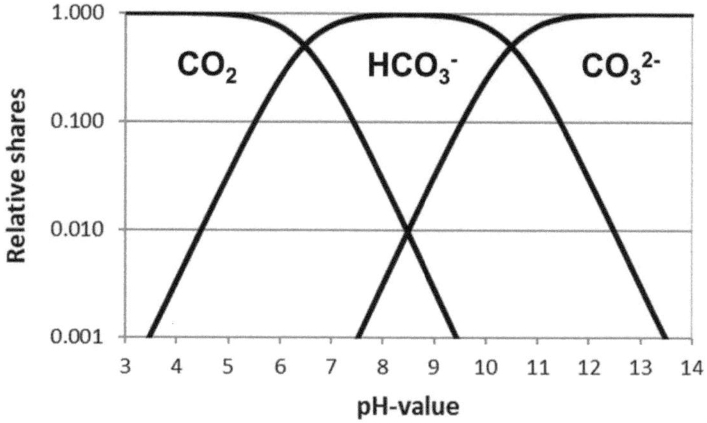

Figure 4.5-3: Relative proportions of carbonic acid forms as a function of pH (10 °C) (I=0), semi-logarithmic plot

If we consider only the pH range 6.5 - 9.5 permissible for drinking water according to the German Drinking Water Ordinance (Figure 4.5-4:), we can see that drinking waters are always so-called "hydrogen carbonate waters", since the HCO_3^- content is predominant and always above 50%.

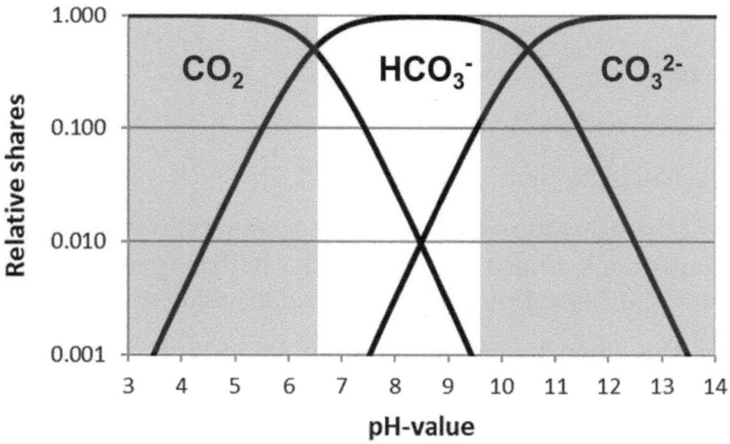

Figure 4.5-4: Relative proportions of carbonic acid forms in the pH range of drinking waters (10 °C) (I=0)

If the formation of calcium and magnesium complexes (chap. 7.6) is taken into account, the distribution of the unbound carbonic acid fractions shown in Figure 4.5-5 results. The values were calculated for a water of medium hardness (80 mg/L Ca and 16 mg/L Mg = 2.7 mmol/L, 15 °dH). It can be seen that complexation affects the equilibria in the carbonic acid system, especially at higher pH values. (Chap. 7.6 and calculation Chap. 10.7.1)

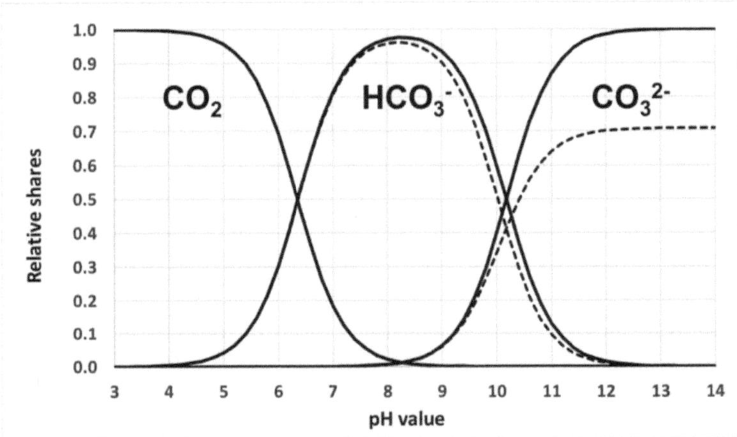

Figure 4.5-5: Proportions of carbonic acid forms as a function of pH, taking into account the formation of Ca and Mg complexes (dashed) (15°dH, I=9.7 mmol/L)

4.6 Buffering

(Buffering basics see chap. 11.6).

(Weak acid, see Acid Strength Basics chap. 11.5)

Since in natural freshwaters carbonic acid is generally the only weak acid in higher concentrations, it also determines the buffering properties (beware of high humic acid concentrations or phosphated drinking waters!).

Examples of the importance of buffering in natural waters and aqueous solutions:

- Effects on **iron corrosion** in water distribution systems
  A high buffering intensity reduces the tendency to corrosion. [54] [55]
- Significance for pH-dependent **biological processes** in natural waters and in water treatment technology. For example, both nitrification and

denitrification lead to pH changes, so that in weakly buffered waters the optimal pH range is easily left and the biological processes only run at a slower rate or come to a standstill:

e.g.	Nitrification	➔ pH reduction	pH optimum : 6 - 8
	Denitrification	➔ pH increase	pH optimum : 7 - 8 [56]

- **Human Blood** is highly buffered between pH 7.35 and 7.45. Below this, the clinical picture acidosis occurs, above it alkalosis. The most important buffer substance is carbonic acid with a concentration of HCO_3^- of 22-26 mmol/L. The partial pressure of CO_2 is 32-45 mmHg (46.6-59.9 hPa or mbar) [57]. The CO_2 concentration in the blood is regulated by gas exchange in the lungs. The partial pressure of CO_2 in the air is significantly lower.

Since carbonic acid is a diprotic acid, there are two dissociation stages with two forms of carbonic acid, each acting as a weak acid:

$$CO_2 + H_2O \rightleftharpoons HCO_3^- + H^+ \qquad pK_1 = 6.47 \qquad (10\ °C) \qquad \text{(Eq 4.6-1)}$$

$$HCO_3^- \rightleftharpoons CO_3^{2-} + H^+ \qquad pK_2 = 10.49 \qquad (10\ °C) \qquad \text{(Eq 4.6-2)}$$

The dissociation of a weak acid can be represented in general terms:

$$AH \rightleftharpoons A^- + H^+$$

The so-called buffer equation, also known as HENDERSON-HASSELBALCH 's equation, is:

$$pH = pK + lg\frac{a(A^-)}{a(AH)} \qquad \text{(derivation see chap. 11.6)} \qquad \text{(Eq 4.6-3)}$$

Buffering of a weak acid AH is relatively greatest when pH = pK, because then the activities of the undissociated acid a(AH) and the acid anion a(A-) are equal, making $lg\frac{a(A^-)}{a(AH)} = 0$.

Thus, two points of maximum buffering can be expected in the carbonic acid system (Table 4.6-1)(see also Figure 4.5-2:).

Table 4.6-1: Buffering maxima and minima of carbonic acid

Maximum buffering			10 °C	25 °C
Maximum 1	:	$pH = pK_1$	6.5	6.4
Maximum 2	:	$pH = pK_2$	10.5	10.3
Minimum	:	$\frac{pK_1 + pK_2}{2}$	8.5	8.4

In Figure 4.6-1: the relative buffering intensity PI is shown, where the PI at pH 10.5 was set as 100%.

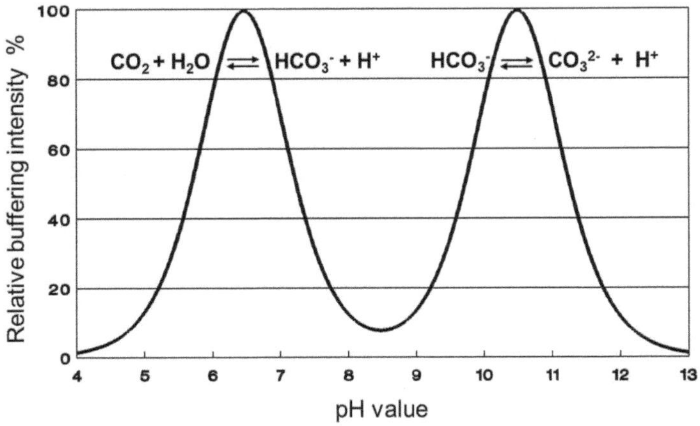

Figure 4.6-1: Relative buffering intensity (PI) of carbonic acid forms in water at 10 °C. c(C) = constant ; t = 10 °C ; I = 0 ; PI at pH 10.5 = 100 %

The absolute value of the buffering intensity depends on the carbonic acid sum c(C) and the pH value. With the carbonic acid sum unchanged, a change in pH can cause large changes in the amount of buffering intensity. This is particularly problematic in soft drinking waters where the pH of calcite saturation is pH 8 or higher, as this provides a minimum buffering intensity (Figure 4.6-2:).

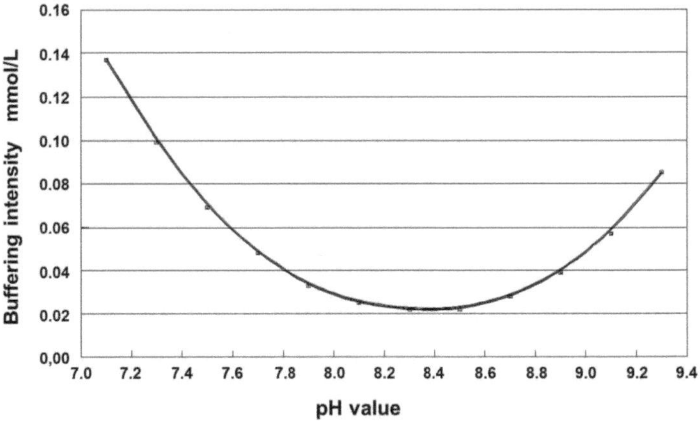

Figure 4.6-2: Calculated buffering intensity of a very soft drinking water (hardness 0.5 mmol/L, 3 °dH) at different pH values. The pH value of the drinking water was 9.3 and was changed for this representation via calculation program.

The definition of the buffering intensity as well as its calculation is discussed in chapter 7.3.

4.7 Solution of lime in water

4.7.1 Preliminary note on lime

Calcium carbonate occurs in nature in three crystallized modifications [1]:

Calcite (calcareous spar)	- hexagonal crystallized
Aragonite (pearls)	- rhombic crystallized
Vaterit	- rhombic crystallized

Aragonite and vaterite are of minor importance. Vaterite forms single crystals and occurs rarely. Aragonite is metastable and passes easily into the stable phase calcite.

If calcium carbonate precipitates as a poorly soluble compound when the solubility product is exceeded from aqueous solutions, the precipitate is initially amorphous and then slowly changes to the crystalline form of calcite. [1]

When heated (burned), calcium carbonate decomposes to form calcium oxide (burned lime) and carbon dioxide:

$$CaCO_3 \longrightarrow CaO + CO_2 \qquad \text{(Eq 4.7-1)}$$

This reaction is used for the production of burnt lime and for the produktion of carbon dioxide.

Calcium oxide reacts with water in a strongly exothermic reaction to form calcium hydroxide = slaking of lime:

$$CaO + H_2O \rightleftharpoons Ca(OH)_2 \qquad \text{(Eq 4.7-2)}$$

Calcium hydroxide is the main component of lime mortar. Calcium carbonate is formed by reaction with the CO_2 in the air, which then leads to the corresponding strength:

$$Ca(OH)_2 + CO_2 \rightleftharpoons CaCO_3 + H_2O \qquad \text{(Eq 4.7-3)}$$

Linguistically, the term "lime" therefore occurs for different chemical compounds:

$CaCO_3$	Calcium carbonate	- Lime
CaO	Calcium oxide	- burnt lime, quicklime
$Ca(OH)_2$	Calcium hydroxide	- Hydrated lime, slaked lime

Common names for naturally occurring calcium carbonate include limestone, chalk, and marble.

In the following sections, the terms lime and calcite are used equivalently.

Lime = Calcite = $CaCO_3$

4.7.2 Lime solution

The dissolution of lime (calcite = $CaCO_3$) in water is the dissolution of a real electrolyte and thus a purely physical process. Ions are dissolved out of the crystal lattice structure of the solid and stabilized by hydration.

The reaction equation for the solution of calcite in water is:

$$\downarrow CaCO_{3(solid)} \overset{K_C}{\rightleftharpoons} Ca^{2+} + CO_3^{2-} \qquad \text{(Eq 4.7-4)}$$

Calcite thus dissolves in water to form calcium and carbonate ions (Eq. 4.7-4). Calcite, however, belongs to the poorly soluble compounds, so that only small amounts of calcium and carbonate ions go into solution.

How much calcite can be dissolved is determined by the solubility product L (see Fundamentals Chap. 11.7):

$$L(CaCO_3) = a(Ca^{2+}) \cdot a(CO_3^{2-}) \qquad \text{(Eq 4.7-5)}$$

Usually, $L(CaCO_3)$ is referred to as K_C : $K_C = L(CaCO_3)$

$$K_C = a(Ca^{2+}) \cdot a(CO_3^{2-}) \qquad \text{(Eq 4.7-6)}$$

The solubility product depends on the temperature of the water.

Example values of constant K_C for different temperatures are given in Table 4.7-1: (calculation according to chap. 7.1). Comprehensive table values can be found at 12.2.

Table 4.7-1: Temperature dependence of solubility product K_C for $CaCO_3$

Temp. °C	K_C (mol/L)^2
0	$4.20 \cdot 10^{-9}$
10	$3.92 \cdot 10^{-9}$
20	$3.52 \cdot 10^{-9}$
25	$3.30 \cdot 10^{-9}$

In Figure 4.7-1: the temperature dependence for K_C is shown graphically. With increasing temperature, the solubility of calcite in water decreases (Figure 4.7-1:).

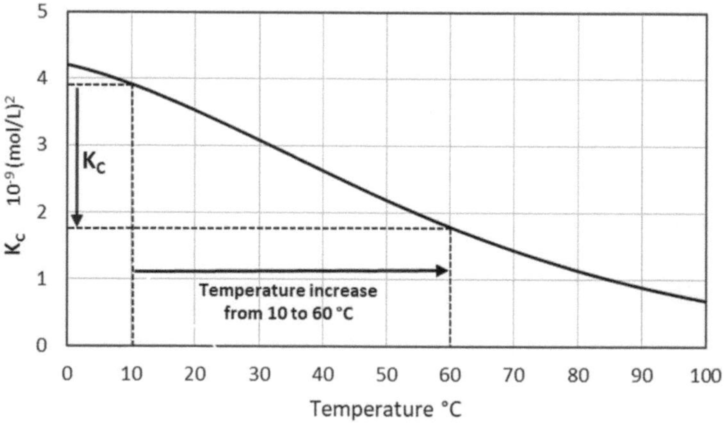

Figure 4.7-1: Temperature dependence of the solubility constant K_C

If the solubility product is reached in a water at low temperature (e.g. 10 °C) and this water is heated, for example for heating purposes (e.g. 60 °C), the solubility product becomes smaller (approx. 50%), the water becomes calciferous, calcite can precipitate (scale formation). If additional CO_2 is outgassed when water is heated in open vessels, the lime scale tendency is increased!

Reaction scheme

When calcite is dissolved in water, carbonate ions are formed (Eq. 1 in Figure 4.7-2:), which react with H^+ in a secondary reaction to form hydrogen carbonate ions (Eq. 2). This leads to an increase in the pH value. We first consider an originally **<u>CO_2-free</u>** system (Figure 4.7-2:).

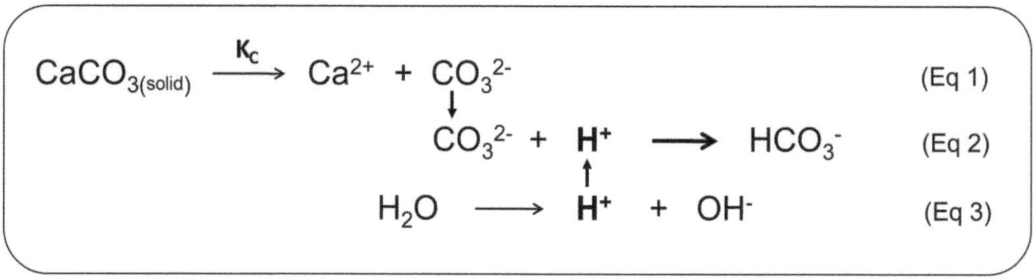

Figure 4.7-2: Reaction scheme of the solution of calcite (CO_2-free system)

The consumption of H^+ ions (Eq. 2) leads to a stronger dissociation of the water (Eq. 3) and to increase the OH concentration, which illustrates the pH increase. Due to the consumption of carbonate ions, equation 1 is shifted to the right, more calcium carbonate dissolves. The reactions proceed until equilibrium is reached (K_C). In very small proportions, CO_2 can also form from hydrogen carbonate.

Annotation: The fact that even low concentrations of CO_2 are formed here should not be considered for the time because of the manageability of the reaction processes.

If **CO_2 is added** to this system from outside, H^+ ions are formed by the reaction of the CO_2 with the water, the water is acidified.

$$CO_2 + H_2O \rightleftharpoons HCO_3^- + H^+ \qquad \text{(KG-3)}$$

This change in pH also shifts all other equilibrium reactions accordingly until the reaction equilibrium is reached again. The entire system of equations in the **reaction equilibrium** is shown in Figure 4.7-3:.

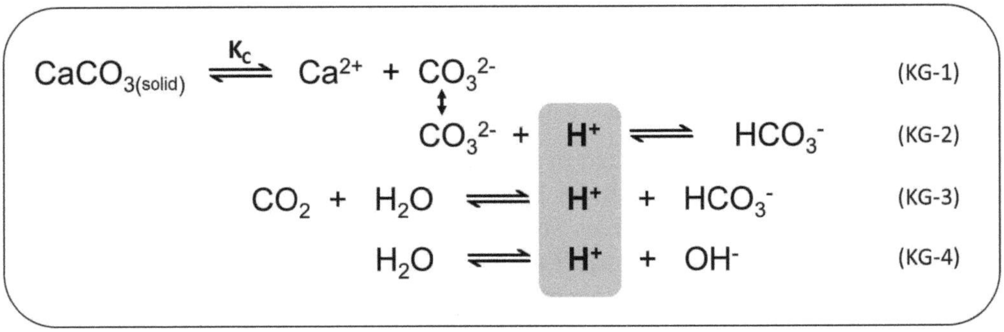

Figure 4.7-3: Reactions in the lime-carbonic acid system

In this system, the concentration of H^+ ions and thus the pH value play the central role, since changes in the H^+ concentration (pH changes) have an effect in **all** reactions. To clarify the relationships, a different representation is therefore chosen for this **lime-carbonic acid system** . The H^+ symbol is, as it were, the central control variable at the heart of the system (Figure 4.7-4:).

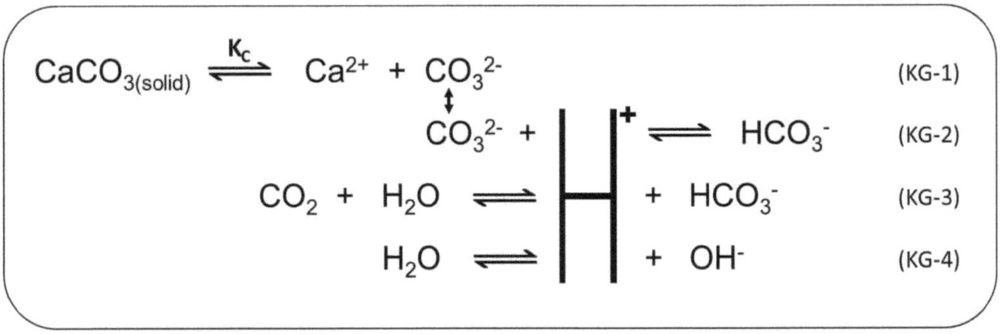

Figure 4.7-4: Reaction diagram of the lime-carbonic acid system in lime-carbonic acid equilibrium

This system of equations illustrates that all reactions are linked via pH.

For all shifts in the carbonic acid system, a specific reaction sequence takes place.

A **primary reaction** changes the H^+ concentration (e.g., by changing the CO_2 concentration). All other reactions react according to their equilibrium in **secondary reactions** to rebalance the change in H^+ concentration.

Nothing changes in the **reaction equilibrium** (Figure 4.7-4:), the so-called **lime-carbonic acid equilibrium**. The water is in the state of calcite saturation. No lime dissolves and no lime precipitates. The pH value prevailing in the reaction equilibrium is called the pH value of calcite saturation (pH_C). The pH_C is specific for each water. It can be measured or calculated.

In balance: $pH = pH_C$

The system of equations shown in Figure 4.7-4: can be applied schematically to all questions of lime-carbonic acid equilibrium.

> <u>Annotation:</u> A closer look at this reaction system may reveal an apparent contradiction, since in reaction (KG-3) formally one HCO_3^- ion is also produced per H^+ ion. This would mean for reaction (KG-2) that the same change occurs on both sides of the reaction equation and thus the reaction equilibrium is maintained. However, it can be shown by converting the mass action equations (example Chap. 10.6) that for the same CO_2 concentration, the HCO_3^- concentration behaves inversely proportional to the H^+ concentration, while the CO_3^{2-} concentration depends inversely proportional to the **square** of the H^+ concentration. This then also leads to a shift in the reaction equilibrium in reaction (KG 2).

For <u>**exemplary application**</u> of this reaction system, we return to the consideration of the change in CO_2 concentration.

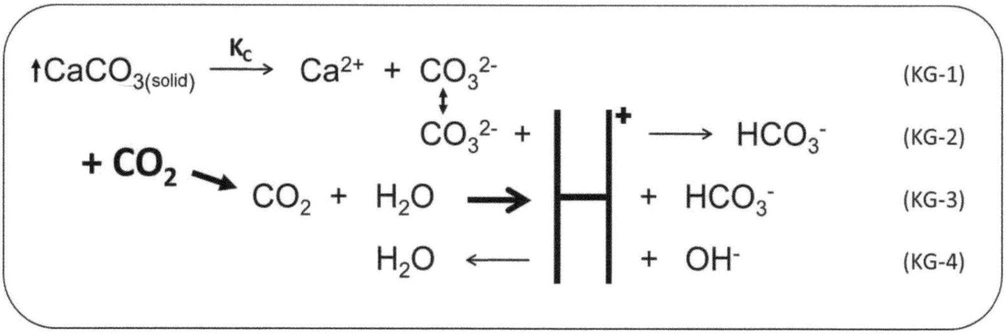

Figure 4.7-5: The lime-carbonic acid system when supplied with CO_2

<u>**Case 1:**</u> **CO_2** is added to the system from the outside.

When CO_2 is added to the system from outside (Figure 4.7-5), H^+ ions (KG-3) are formed by the reaction of the CO_2 with the water, and the water is acidified: **Primary reaction : H^+ formation.**

Due to their equilibrium position, the other reactions involved try to compensate for the increase in H^+ concentration: **Secondary reactions : H^+ consumption.**

<u>Primary reaction:</u>
KG-3 reacts by increasing the CO_2 concentration to the right: **H^+ formation**

<u>Secondary reactions:</u>
KG-4 reacts to the left, and OH^- form water (neutralization reaction)
KG-2 reacts to the right, CO_3^{2-} is consumed !
KG-1 reacts to the right as CO_3^{2-} concentration decreases

➔ Lime is dissolved !

➔ So if CO_2 is present in the water, more lime can dissolve.
The richer the water is in CO_2, the more lime dissolves.

The reactions continue until the system is back in equilibrium. The equilibrium is reached with an overall increased H^+ concentration - the pH value decreases !

The amount of CO_2 present after the reaction equilibrium has been set, which is capable of "holding in solution" a certain amount of lime, was historically referred to as "equilibrium carbonic acid." This term is outdated and no longer useful today, but can serve to understand older texts.

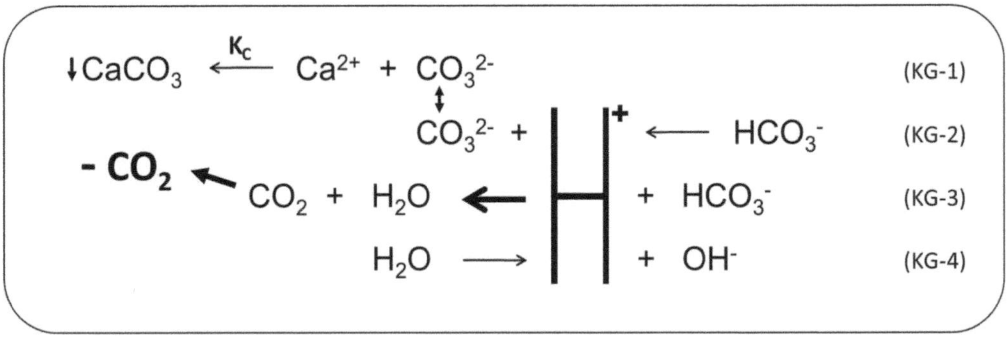

Figure 4.7-6: The lime-carbonic acid system when CO_2 is withdrawn

<u>**Case 2:**</u> **CO_2** is **removed** from the system.

If the CO_2 (or a partial amount) is removed again (Figure 4.7-6:), the reactions take place in the opposite direction. By increasing the CO_3^{2-} concentration, the solubility product is exceeded and lime can precipitate!

The reactions in Figure 4.7-6 proceed as follows:

Primary reaction:

KG-3 reacts by decreasing the CO_2 concentration to the left: H^+ consumption

Secondary reactions:

KG-4 reacts to the right, H^+ and OH^- are formed
KG-2 reacts to the left, CO_3^{2-} is formed !
KG-1 reacts to the left as CO_3^{2-} concentration increases

➔ the solubility product is exceeded, lime can precipitate

Technically, CO_2 removal is performed to physically deacidify a water. In the case of deacidification processes, the lime-carbonic acid equilibrium scheme can help to understand why different final pH values are obtained depending on the process technology (see chapter 9.1.6).

The reaction equilibrium of the described equilibrium reactions in the lime-carbonic acid system does not occur in natural waters in every case. In the case of groundwater, the contact time with the soil is often insufficient or the soil contains only small amounts of lime (sands, low-lime rock), so that the water predominantly still has a lime-dissolving tendency.

If a water in calcareous soil (e.g. sand-lime brick, dolomite) reaches the lime-carbonic acid equilibrium, this cannot be exceeded under constant external conditions (pure dissolving process up to the solubility limit!). If external conditions change, for example due to CO_2 loss in source areas, such water may acquire lime-separating properties (chap. 5.5).

If a water with a tendency to dissolve lime ("aggressive") is pumped for drinking water supply purposes, there is usually no solid $CaCO_3$ in the pipes of water extraction and distribution systems, so the equilibrium remains disturbed. However, if this water comes back into contact with materials containing $CaCO_3$ (e.g. concrete, fiber cement, cement mortar linings of steel pipes), lime can dissolve and the contact material is attacked.

With regard to the limescale dissolving capacity of drinking water, there are limit values in the German Drinking Water Ordinance (chapter 9.1.1). If these values cannot be maintained, the water must be deacidified (chapter 9.1).

5 The importance of CO_2 in the water cycle

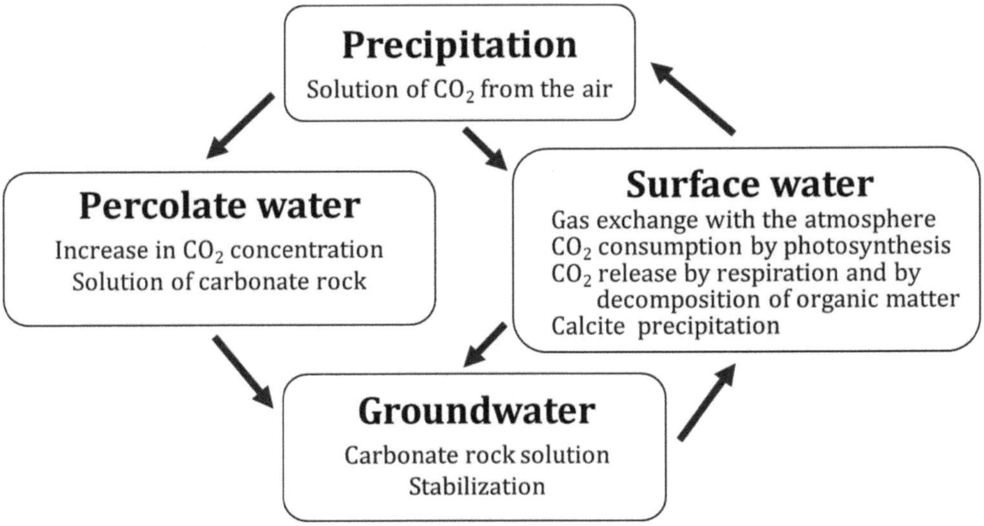

Figure 4.7-1: CO_2 in the Water cycle - overview

5.1 Weathering

Weathering leads to natural decomposition of rock. The CO_2 dissolved in the water or the carbonic acid formed contributes to weathering. Effective here is the formation and provision of protons H^+ or also the buffering of pH changes.

Only weathering processes involving carbonates or in which the carbonic acid system has a part will be considered here.

Lime will serve as the main example for carbonates here. The solution of lime (weathering by solution) is initially independent of CO_2 and carbonic acid.

Lime dissolves in water to form Ca^{2+} and carbonate ions. The solubility is limited and is defined by the solubility product $L = K_C$ (chapter 4.7.2).

$$CaCO_3 \rightleftharpoons Ca^{2+} + CO_3^{2-} \qquad \text{(Eq 5.1-1)}$$

Since the carbonate ion is part of the carbonic acid system, the CO_2 concentration affects the solubility of lime via the equilibrium reactions of the carbonic acid system and the pH. High CO_2 concentrations promote the dissolution of lime and other carbonates. This reaction is often given as a summation equation. This is stoichiometrically correct, but the actual reaction sequence is different (for the exact reaction sequence, see section 4.7).

Summarized (simplified !!):

$$CaCO_3 + CO_2 + H_2O \leftrightharpoons Ca^{2+} + 2\ HCO_3^- \qquad \text{(Eq 5.1-2)}$$

The dissolution of carbonates and the associated release of carbonate ions increases the pH value. This cannot be derived from the summary reaction equation (see chap. 4.7).

Solution of lime in limestone can lead to karstification.

Silicate weathering (hydrolysis) can also be promoted by carbon dioxide and carbonic acid. The weathering of silicate minerals is referred to as an incongruent solution process [58] this is because only part of the weathering molecule goes into solution and another part transforms into a secondary mineral. During this weathering process OH^- ions are released.

As an example this is shown here for potassium feldspar [43] :

$$\underset{\text{(Potassium feldspar)}}{2KAlSi_3O_8} + 11H_2O \rightarrow \underset{\text{(Kaolinite)}}{Al_2Si_2O_5(OH)_4} + \underset{\text{(Silica)}}{4H_4SiO_4} + 2K^+ + 2OH^- \qquad \text{(Eq 5.1-3)}$$

CO_2 and carbonic acid promote this hydrolytic weathering, since the released OH^- ions are buffered via the carbonic acid system:

Summarized (simplified!):

$$CO_2 + OH^- \rightleftharpoons HCO_3^- \qquad \text{(Eq 5.1.-4)}$$

The exact reaction sequence is shown in chapter 4.7 .

The kaolinite formed belongs to the phyllosilicates (clay).

5.2 Rainwater

Raindrops have intensive contact (large surface area) with the air and CO_2 dissolves from the air in the rainwater according to the laws for the dissolution of gases in water (chapter 4.1). Rainwater saturated with carbon dioxide without impurities has a pH value of 5.56 (calculation example 10.5).

The rainwater hits the surface of the ground. Some of the water drains away superficially and remains in contact with the atmosphere. The other part of the rainwater seeps into the ground and leads to the formation of new groundwater.

5.3 Leachate

The organic substances in the humus layer of the soil are broken down by the bacteria living there. In the process, O_2 is consumed and CO_2 is produced.

Mineralization:

Bacteria
↓
Organ. Substance + O_2 → CO_2 + Minerals + **hardly degradable residue**
↓ **= humic substances !**
↓
Solution in leachate

Seepage seeps vertically into the subsurface through the gaps of the soil . In the process, contact with the atmosphere is lost and the dissolution of CO_2 then depends on the composition of the soil air. In soil air, CO_2 is present in higher partial pressures (Table 4.1-2).

The top layer of soil is usually enriched with humus substances. Here, intensive biodegradation of organic matter takes place, releasing CO_2. As the gas exchange between soil air and atmosphere decreases with increasing depth, CO_2 accumulates in the soil air. As a result, more CO_2 dissolves in the leachate, corresponding to the increasing partial pressure (Table 4.1-2). The seepage then passes into the groundwater body, which backs up over impermeable strata and exhibits horizontal flow.

5.4 Groundwater

Already on the seepage path, but also in the area of the groundwater body (aquifer), reactions of the water and its constituents with the soil body occur. Among the water constituents, the forms of carbonic acid are of particular importance here.

The increased CO_2 solution in the seepage section (Figure 5.4-1) leads to an acidification of the water (pH reduction). This pH shift is at least partially compensated by ion exchange processes and/or by solution processes:

- e.g. calcite (lime) or other carbonates are dissolved
- e.g. conversion of silicate minerals

If the residence time of the groundwater in the soil is long enough, the water stabilizes, i.e. dissolution and precipitation processes are in equilibrium, and the constituents of the water no longer change.

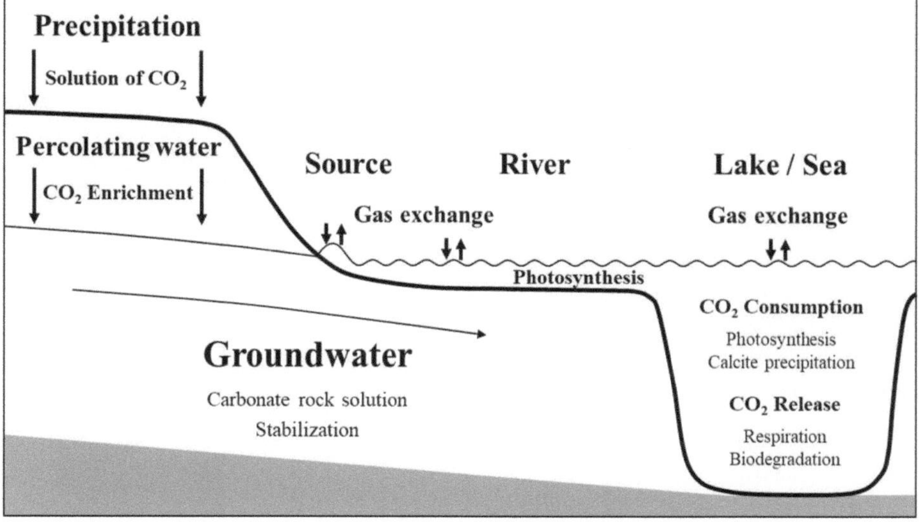

Figure 5.4-1: Carbon dioxide in the water cycle

This state of equilibrium with respect to calcite and CO_2 then corresponds to the so-called lime-carbonic acid equilibrium.

If the stabilization is not completed, this water still has a lime solvent tendency.

5.5 Spring water

Groundwater may come to the surface in the form of a spring . From that moment on, the CO_2 partial pressure of the air is again decisive for the solubility of the CO_2 in the water. Since as a rule in the spring water more CO_2 is dissolved than corresponds to the solution equilibrium with the air, CO_2 outgasses. This leads to a shift in the carbonic acid balance in the water and the pH value of the spring water increases (Figure 5.6-1)!

If the groundwater was previously lime-dissolving, the pH increase causes the lime-dissolving tendency to decrease. The lime-carbonic acid equilibrium may be reached or it may even eventually be exceeded and lime precipitation may occur.

If the groundwater was already in lime-carbonic acid equilibrium, the outgassing of CO_2 and the associated increase in pH now exceeds the solubility for lime and precipitation of lime (calcite) can occur. So-called swelling limestones then frequently form (Figure 5.5-1).

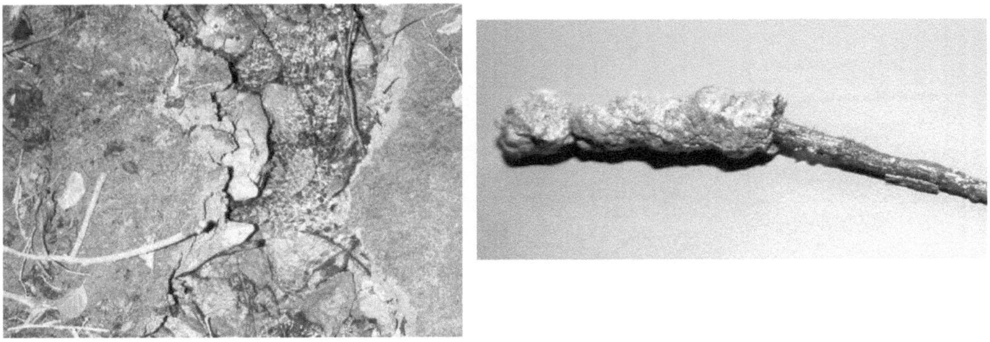

Figure 5.5-1: Lime deposits in a spring creek and on a piece of wood from this creek

5.6 Running waters

In the source area, the pH value initially increases due to CO_2 outgassing (see above). Spring mosses, algae, and higher aquatic plants photosynthesize when exposed to light. To do this, they absorb CO_2 from the water. This additional CO_2 removal results in a further increase of the pH value.

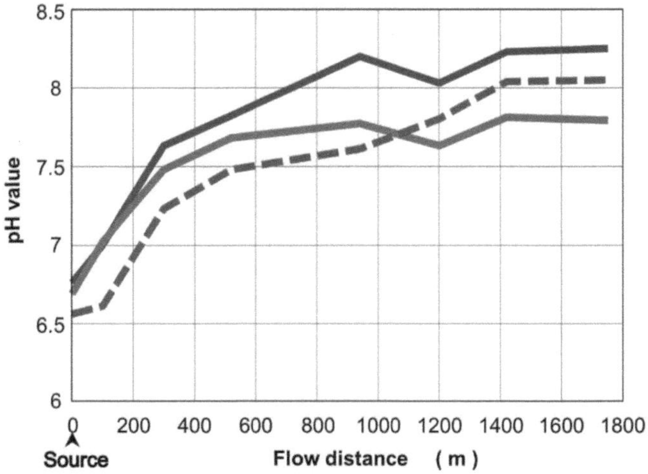

Figure 5.6-1: pH values of a spring creek in flow direction (different years, each in May, noon)

Due to the CO_2 discharge in the source area and the photosynthesis processes, the pH value in a spring creek increases during the day in the direction of flow (Figure 5.6-1)!

The diurnal variation of pH (Figure 5.6-2) shows an increase in pH in the morning due to CO_2 removal by photosynthesis. With decreasing solar radiation and thus reduced CO_2 consumption, the pH value decreases again in the afternoon hours.

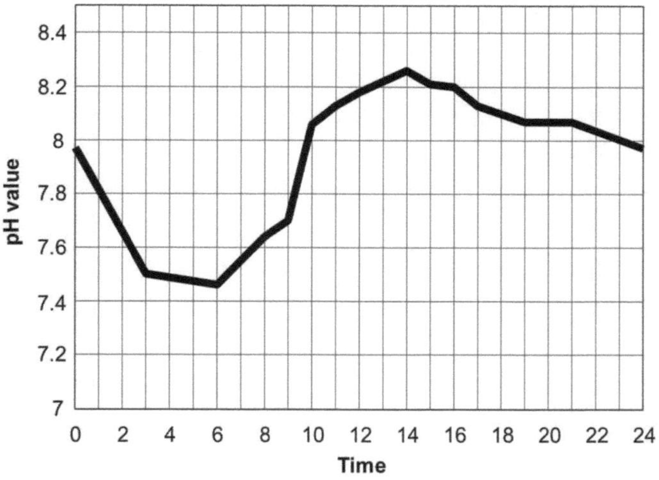

Figure 5.6-2: pH values of a spring creek in the diurnal cycle (Measuring point 1400 m after source)

During the night hours, CO_2 production by biological respiration and by degradation processes predominates, and the pH value continues to drop. With the onset of sunlight in the morning and thus the start of photosynthesis, more CO_2 is then again consumed than produced, so that the pH value rises again.

The gas exchange with the air is very intensive in flowing waters due to turbulence, so that in smaller, very turbulent waters the CO_2 in the water can be in solution equilibrium with the air due to the intensive exchange processes. The **natural** pH value of calcite saturation (pH_C) in such a water body is about 8.2 (Figure 5.6-1, Chapter 6).

In the case of aquatic plants, a general distinction is made between three types of plants, depending on the type of usable carbon source (according to RUTTNER from [59]).

The preferred carbon source of plants can be explained in each case as the existing adaptation to the particular habitat.

<u>Plant types with respect to carbon source in water:</u>

1. Fontinalis type (Fontinalis = spring moss)
   ➔ Only dissolved CO_2 is absorbed.

2. Elodea type (Elodea = water plant)
   ➔ Besides CO_2, HCO_3^- is also used, but CO_2 can be absorbed about 5 times faster. However, since only CO_2 can be bound during photosynthesis, the HCO_3^- must be split:

$$HCO_3^- \rightarrow CO_2 + OH^- \qquad \text{(Eq 5.6-1)}$$

3. Scenedesmus type (Scenedesmus = planktonic green algae)
   ➔ Predominantly uptake of HCO_3^-, hardly any CO_2 (ratio about 25 : 1)

In the source area of a running water, CO_2 is usually present in excess (fontinalis type). In small ponds and in the shore area of lakes (littoral), both CO_2 and hydrogen carbonate can already be expected in larger quantities (Elodea type). In the open water zone of lakes (Pelagial), hydrogen carbonate is the predominant carbonic acid form due to the higher pH, so planktonic algae (Scenedesmus type) are adapted to these conditions.

5.7 Lakes

On the exchange processes in a lake the summer temperature stratification of the water (summer stagnation) has a great influence.

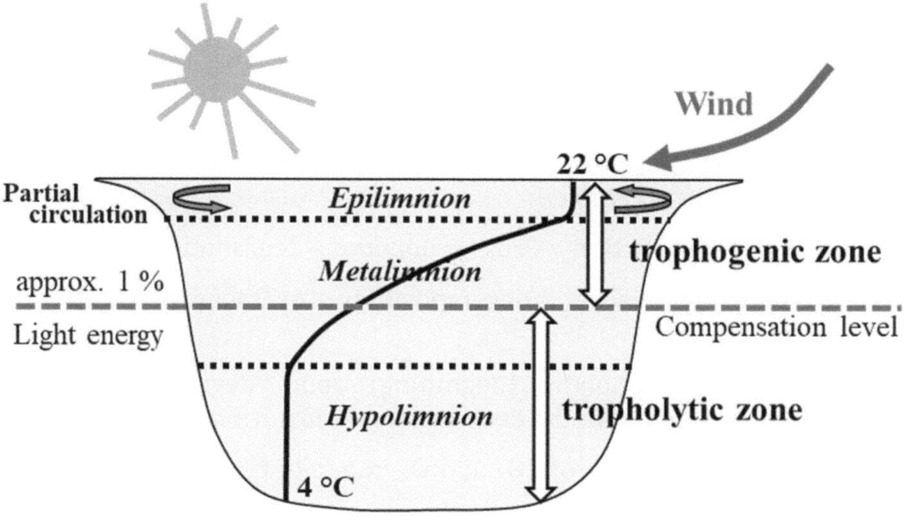

Figure 5.7-1: Zoning in a lake during summer stagnation

The stratification isolates the deep water and the exchange processes are limited to the epilimnion (Figure 5.7-1:).Gas exchange with the atmosphere takes place only at the lake surface, more or less intensively, depending on turbulence.

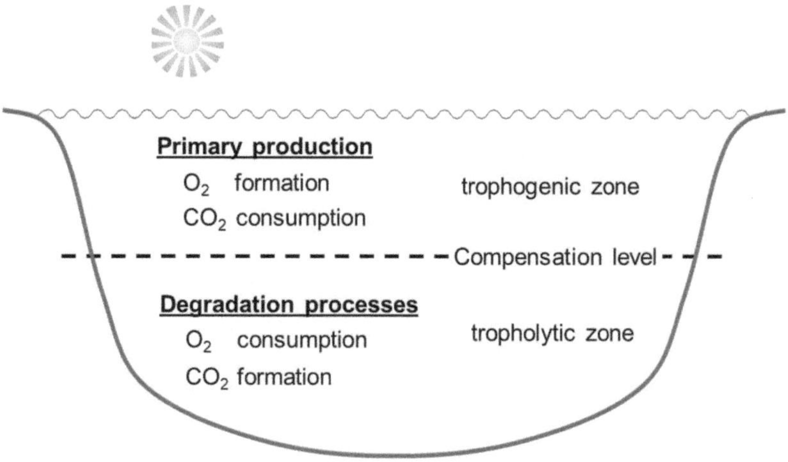

Figure 5.7-2: Schematic representation of the relationship between oxygen and carbon dioxide in a lake

However, the processes of photosynthetic primary production and decomposition (dissimilation) in the lake body are decisive for the CO_2 content of the water of a lake.

In the light-flooded trophogenen (building up) zone (Figure 5.7-2), CO_2 is consumed and O_2 is produced by photosynthesis. The CO_2 concentration decreases and the pH value increases.

Compensation level indicates the depth of a water body below which net production is no longer possible.

Net production = Gross production - degradation

In this area, photosynthesis and decomposition processes balance each other out.

In the underlying tropholytic (degrading) zone, degradation processes predominate, enriching CO_2 in the deep water and lowering the pH.

Therefore, the depth profiles of CO_2 and O_2 in a lake typically run in opposite directions (Figure 5.7-3:).

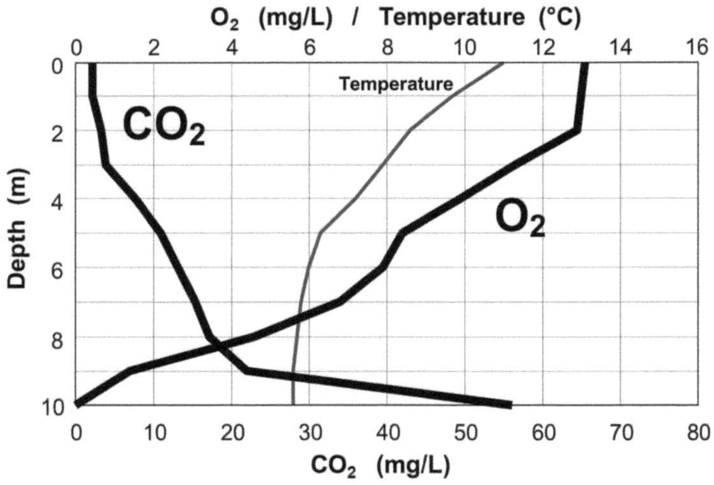

Figure 5.7-3: Opposing depth profiles of the measured values of CO_2 and O_2 (eutrophic lake, summer stagnation, 10 m deep)

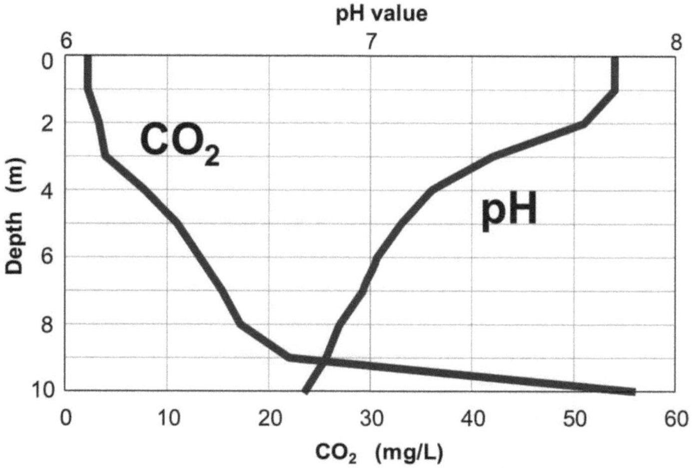

Figure 5.7-4: Opposing depth profiles of the measured values of CO_2 and pH value (eutrophic lake, summer stagnation, 10 m deep)

The lake water's content of CO_2 determines the pH value of the water. One can also see here the opposite course of the depth profile in Figure 5.7-4. The pH scale is widely spread for clarity.

In summer, the deep water of a lake is isolated by the stable thermal stratification of the water (summer stagnation). In spring and autumn, in isothermal conditions (same temperature at approx. 4°C), surface water can reach depths through circulation processes (wind).

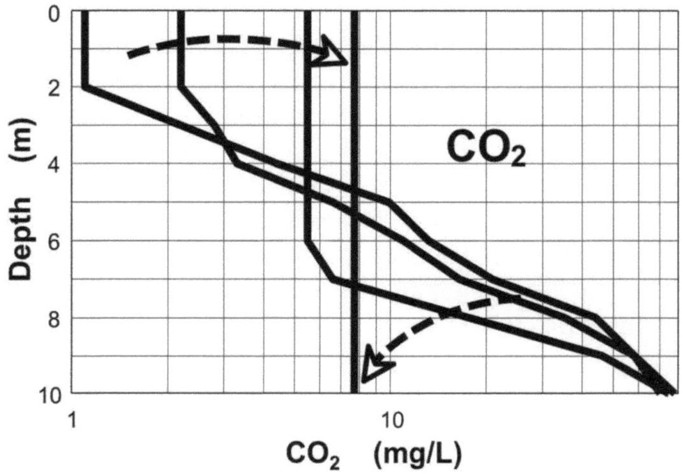

Figure 5.7-5: Concentration balance of CO_2 in the depth profile during full circulation (semi-logarithmic representation) (eutrophic lake, summer stagnation, 10 m deep)

Full circulation then occurs in each case, resulting in complete mixing of the water layers and equalization of the concentrations of the water constituents. When equalizing the concentrations these are again opposite processes for CO_2 and O_2 (Figure 5.7-5: and Figure 5.7-6:).

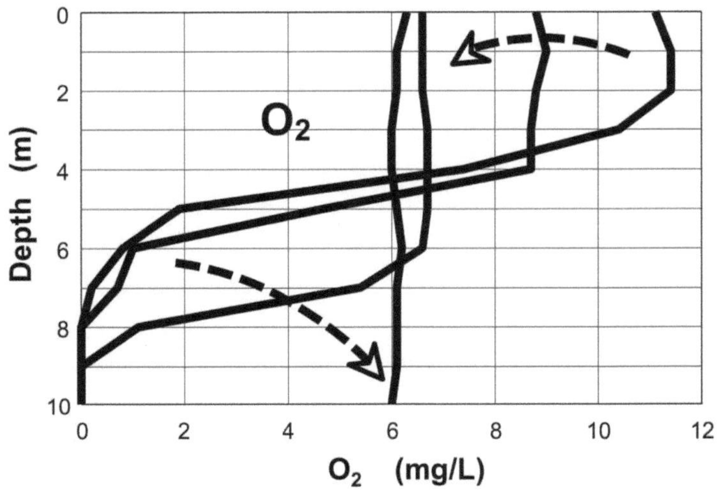

Figure 5.7-6: Concentration balance of O_2 in the depth profile at full circulation (eutrophic lake, summer stagnation, 10 m deep)

In summer, especially in nutrient-rich (eutrophic) lakes, intensive photosynthesis and formation of organic biomass (primary production) occurs in the surface layer (epilimnion). HCO_3^- is taken up as carbon source (Scenedesmus type) by planktonic algae, which account for most of the primary production in lakes (chap. 5.6). The pH value of the water increases. The extent of the pH increase depends on the buffering capacity of the water.

If the water is relatively soft with low buffering intensity, the pH value can rise to above 10 during intense photosynthesis (Figure 5.7-7:). The increase in pH leads to a shift of the carbonic acid proportions towards carbonate (section 4.5), which cannot be used by the algae. There is a depletion of HCO_3^- in the surface water due to this shift, which may limit primary production due to nutrient deficiency (CO_2).

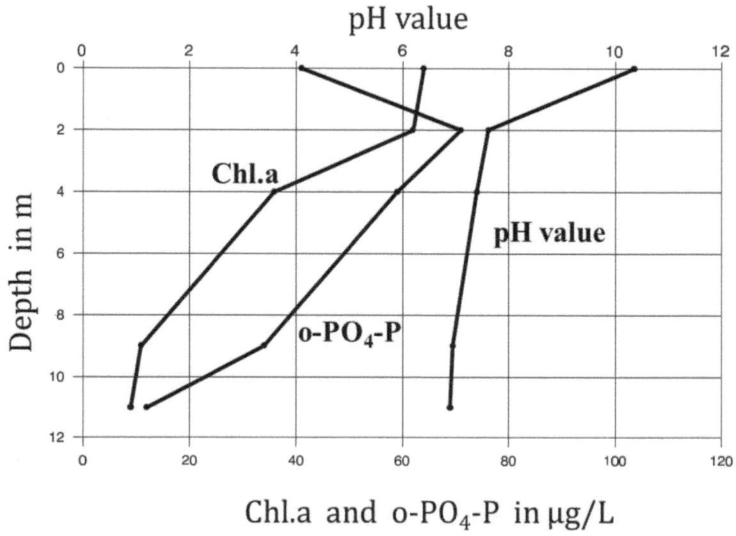

Figure 5.7-7: Example of a eutrophic reservoir, weakly buffered water
Summer stagnation

The increase in pH due to intensive photosynthetic processes often causes the solubility product of lime to be exceeded, resulting in calcite precipitation, which is particularly visible in the littoral zone (littoral) by a white calcite coating on the plants. In the open water (pelagic) the precipitation products sink, but they can dissolve again in the CO_2-rich deep water. (Figure 5.7-8:)

5.7.1 Carbon cycle in standing water

The carbon cycle in water bodies begins with carbon dioxide being bound by photosynthesis in the form of organic matter. CO_2 is then released again via respiration of the organisms and via decomposition processes. (Figure 5.7-8:)

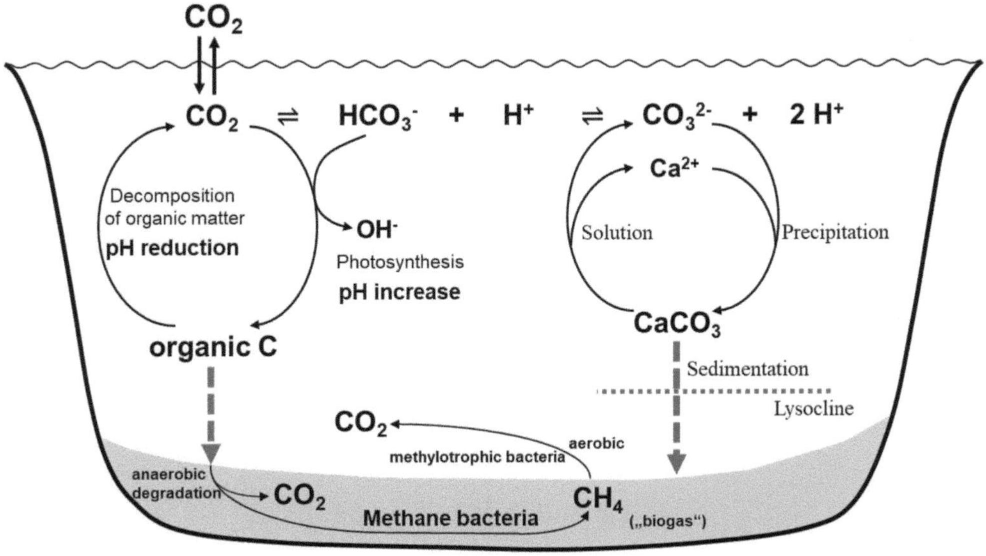

Figure 5.7-8: The carbon cycle in lakes

Some of the organic matter produced in the water body sediments. Microbiological degradation processes start in dead material already during sinking. Residual degradation occurs in the sediment. However, since anaerobic conditions are partly present there, further decomposition of the organic matter can proceed via fermentation processes in which primarily organic acids, alcohols, CO_2 and H_2 are formed. These substances can eventually be used by methane bacteria (e.g., Methanobacterium, Methanococcus) [60] further converted to methane .

Aerobically (in the presence of air), the methane can then finally be oxidized back to CO_2 by methylotropic bacteria (e.g., Methylomonas, Methylococcus, Methylosinus) [60].

Annotation:

Methane occurs in lake sediments and marshes (marshgas), but also in septic tanks of sewage treatment plants and in the rumen of ruminants.

The release of methane, a greenhouse gas, plays a significant role, particularly in the case of overcropped rice cultivation and factory farming (ruminants).

If the pH in the open water increases too much due to CO_2 removal by photosynthesis, the carbonic acid system is shifted towards CO_3^{2-}. The solubility product of lime may be exceeded and lime may precipitate. Sedimentation of the lime particles occurs. In deep water, high CO_2 concentrations below the so-called lysocline can cause partial dissolution of the lime.

5.8 World Oceans

In principle, the oceans behave in the same way as the lakes.
Atmospheric CO_2 dissolves in the surface layer of seawater. Compared to freshwater, the solubility is slightly lower due to the salinity of 3.5% (35‰) (chapter 4.1).

	Freshwater	Seawater	
10 °C	0,99	0,82	mg/L CO_2

Due to photosynthetic activity, CO_2 is bound in organic biomass in the light-flooded zone.

With incompletely degraded biomass, organically bound carbon constantly sediments into the deep water, where it is converted to CO_2 by biodegradation or is incorporated into the sediments as an organic component.

Carbonates are additionally removed from the free water by the formation and deposition of calcareous skeletons (esp. foraminifera, calcareous algae and corals). A part of the lime particles and lime skeletons dissolves again below the so-called lysocline, but the other part leads to the formation of lime sediments and lime rocks, which only slowly return to the total carbon cycle in geological periods via mountain building and weathering (chap. 3.1.2).

The world's oceans, with their large exchange area of 67 % of the earth's surface and their large volume and long exchange times, form a huge CO_2 reservoir and thus have a balancing effect on the CO_2 content of the atmosphere. Minor disturbances of the CO_2 content of the atmosphere are briefly trapped in the well-mixed surface layer, which, however, has only a limited capacity. The deep water, on the other hand, has a very large capacity, but can only react slowly due to the low exchange by the ocean currents. Acidification the world's oceans has no effect on the **physical solubility** of CO_2 in seawater. This depends only on the pressure and temperature. But the **absorption capacity** for CO_2 is lowered by the pH reduction. The decreasing pH leads to a shift of the carbonic acid balance towards CO_2, so that the CO_2 concentration in the water increases and thus less can be absorbed from the air.

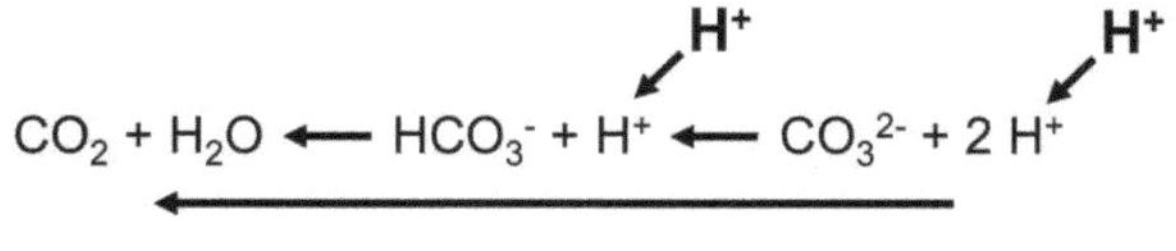

Figure 5.8-1: Shift of the carbonic acid system due to acidification

However, warming of the surface water leads to a decrease in the physical solubility of CO_2 in the surface layer.

Global temperature rise therefore reduces the uptake of CO_2 in the surface layer of the oceans by two processes:

- Decrease in physical solubility due to temperature increase.
- Decrease in CO_2 uptake capacity due to pH decrease.

5.9 Raw water for drinking water treatment

Groundwater

Groundwater, which is pumped for drinking water treatment, is transported in closed pipes under pressure without contact with the ambient air. The composition of the water remains as it was in the aquifer.

If the groundwater have lime-dissolving properties, it must usually be deacidified, since on the one hand corrosion processes must be minimized and on the other hand the adjustment of the lime-carbonic acid balance is required for drinking water in certain areas according to the German Drinking Water Ordinance (chap. 9.1.1).

Spring water

Spring water, which is obtained in a spring tapping, is emergent groundwater and therefore has essentially the characteristics of groundwater.

Surface water

In Germany, surface water, reservoir water, lake water (e.g. Lake Constance) or river water (e.g. Danube) is used directly for drinking water treatment. In standing waters, the raw water is always drawn from the cold deep layer (hypolimnion). In deep water, CO_2 has accumulated due to biological degradation processes, so that these waters usually have lime-dissolving properties and thus have to be deacidified.

If running waters are used as raw water, enriched CO_2, especially at night (chap. 5.6), and thus lime-dissolving properties, can be expected here, depending on the size of the watercourse, contamination and turbulence of the runnig water.
Here, too, deacidification of the water would then be necessary.

5.10 Cold water geysers

Geysers are fountains where groundwater is periodically ejected in the form of a fountain. In hot water geysers, the heating of water to boiling and the formation of steam by magma at not too great depth is what triggers the geysers to jump.

It is different with the cold water geyser. There, the CO_2 dissolved in the water is what drives the geyser to jump. An interaction of gas solubility and water column pressure triggers the geyser to jump. The intervals between eruptions depend on the specific geological conditions. In Andernach on the Rhine, the highest cold-water geyser in the world was drilled with a jump height of up to 60 m with an interval of about 2 hours [61].

Figure 5.10-1: Cold water geyser Andernach (Germany): - escaping gas bubbles, - developing, - full eruption

The prerequisite for a cold-water geyser is the release of gaseous CO_2 from magma deposits in deep layers of the earth and a development through a borehole. After an eruption, the geyser is largely empty and is then refilled to the pressure-induced filling level by groundwater flowing in from the sides underground. When CO_2 escapes from fissures and cracks in the rock in the deep layers of the earth and encounters groundwater, it dissolves in the water until saturation occurs. Since the partial pressure of CO_2 at depth is very high and the total pressure is also very high due to the water column above, large amounts of CO_2 can dissolve in the water.

However, further escaping gas can no longer dissolve in the gas-saturated water and gas bubbles rise in the water column of the well pipe.

Two processes are now overlapping here. The higher the gas bubble rises in the water column, the more the pressure in the column decreases and the gas bubble increases its volume and displaces more water.

The increase in volume and the associated displacement of water by the gas bubbles leads to an increase in the water column up to the overflow at the well head. The overflow reduces the weight of the water column and thus also the pressure at depth. Now the second process is getting underway. The reduced pressure decreases the solubility of the gas in the water, resulting in supersaturated water from which new gas bubbles can outgas. These newly formed gas bubbles also rise and expand in volume. So much CO_2 is released that large coherent gas bubbles form, filling the entire cross-section and drive the water column upward, leading to the eruption of the geyser.

Since the solubility of CO_2 in water is linearly dependent on pressure, decreasing pressure decreases solubility. In the Andernach example, with a pressure of about 35 bar at a depth of 350 m, the gas solubility drops to about a 35'tel when the water rises and the resulting pressure drop to the surface (about 1 bar).

$c_w \sim p$ $\qquad\qquad$ $V \sim 1/p$

c_w concentration in water
p pressure
V Volume

The gas volume increases with decreasing pressure and there is about a 35-fold increase in volume from about 35 bar at depth to ambient pressure as the water column emerges at the surface.

5.11 Hardly soluble carbonates

Hardly soluble carbonates are of great importance in water chemical processes.

Table 5.11-1: Overview of the solubility products of hardly soluble carbonates

Compound	L in $(mol/L)^2$ or $(mol/L)^5$	Temp. °C	Source
$CaCO_3$	$3.30 \cdot 10^{-9}$	25	[52]
$FeCO_3$	$1.99 \cdot 10^{-11}$	25	[44]
$MgCO_3$	$0.65 \cdot 10^{-5}$	25	[44]
$MnCO_3$	$3.98 \cdot 10^{-11}$	25	[44]
$PbCO_3$	$7.76 \cdot 10^{-14}$	25	[44]
$Pb_3(OH)_2(CO_3)_2$	$1.66 \cdot 10^{-47}$	25	[44]

Calcium carbonate $CaCO_3$

The solubility of $CaCO_3$ is discussed in the separate chapters on lime solution and lime-carbonic acid equilibrium (chapter 4.7).

Calcareous deposits in the primordial oceans led to the formation of thick calcareous sediments, which then partly came to the earth's surface through tectonics (e.g. Dolomites). The lime deposits present today are sedimentary rocks and were formed by lime precipitation as well as sedimented lime shells from small marine organisms.

Lead carbonate $PbCO_3$

For a long time, lead pipes were installed in house installations and in house connection pipes for drinking water supply. They had the great advantage of being easy to lay. But although it was already known to the Romans that lead was harmful to health, there were differing views among experts on the potential hazards posed by lead pipes. Only the DIN 2000 [62] expresses itself unambiguously in 1973: "Lead pipes should therefore not be used for new drinking water pipes". Even today, not all lead service lines have been replaced, although intensive work is underway.

Particularly high values in lead pipes still installed are found in drinking water after stagnation periods in the domestic installation area. The solubility of the corrosion products formed, especially lead carbonate $PbCO_3$ and basic lead carbonate $Pb_3(OH)_2(CO_3)_2$, is decisive for the level of lead concentration. In theoretical considerations, one must take into account the additional formation of various lead hydroxo and lead carbonate complexes, such as $PbCO_3^0$ and $Pb(CO_3)_2^{2-}$, to obtain realistic values.

Iron-II-carbonate $FeCO_3$ (Siderite)

Iron content of groundwaters

The solution of iron-III compounds is characterized by the low solubility product of $Fe(OH)_3$ ($L=1.0 \cdot 10^{-38}$ $(mol/L)^4$) [63] so that Fe^{3+} ions can only occur in traces in oxygenated water. However, in the absence of oxygen, reduction of trivalent iron to divalent iron occurs in reduced aquifers. Fe-II compounds are much more soluble, so high concentrations of Fe^{2+} can occur in such waters. However, even here there is ultimately a limit to the solubility, and that is $FeCO_3$, which has a low solubility with a solubility product of $L = 1.99 \cdot 10^{-11}$ $(mol/L)^2$.

Iron corrosion

When the primary corrosion product Fe^{2+} can react to form the poorly soluble $FeCO_3$ (siderite), well protective top layers form. The corrosion rate is reduced. Whether siderite forms or direct iron oxidation occurs depends on water conditions, especially pH and hydrogen carbonate concentration. (Siderite model according to SONTHEIMER [49]).

6 Thought model for carbonic acid in water

6.1 Open system

When considering an open system, **gas exchange** with the ambient air is always possible. As a prerequisite it should be considered that with regard to the gas solution an **equilibrium adjustment always** takes place

These conditions are most likely to be encountered in nature in raindrops, and to some extent can also be expected in very turbulent flowing smaller streams where the gas exchange with the atmosphere is still very large.

Open system - always gas exchange with air and equilibrium adjustment

$$\downarrow\uparrow$$

$$CO_2 + H_2O \rightleftharpoons HCO_3^- + H^+ \rightleftharpoons CO_3^{2-} + 2H^+$$

Figure 6.1-1: Reaction equation for the open carbonic acid system

The concentration of CO_2 dissolved in gaseous form in the water remains **constant**, since the solution equilibrium with the air can always be re-established even in the event of shifts in the carbonic acid system caused by pH changes. Thus, the same concentration is always present in the water (c_W):

$$c_W = c(CO_2) = \textbf{constant} \qquad \text{(Eq 6.1-1)}$$

If one changes the pH value in this system, the proportions of the carbonic acid forms change in dependence on it, whereby the concentration of the CO_2 does not change due to the always occurring equilibrium adjustment in the gas solution, which depends only on the temperature and on the CO_2 concentration in the air (the ionic strength remains unconsidered for simplification).

CO_2 is present as a certain fraction $x(CO_2)$ of the carbonic acid sum $c(C)$, which can be calculated via the dissociation constants and depends on the pH value. The same applies to HCO_3^- ($x(HCO_3^-)$) and CO_3^{2-} ($x(CO_3^{2-})$) (Chapter 4.5).

Since the concentration $c(CO_2)$ is constant and known (see calculation example 0), the carbonic acid sum $c(C)$ can be calculated via $x(CO_2)$ as a function of the dissolved CO_2 :

$$c(CO_2) = x(CO_2) \cdot c(C) \quad \rightarrow \quad c(C) = \frac{c(CO_2)}{x(CO_2)} \qquad \text{(Eq 6.1-2)}$$

From c(C), the concentrations $c(HCO_3^-)$ and $c(CO_3^{2-})$ can then in turn be calculated via $x(HCO_3^-)$ and $x(CO_3^{2-})$.

$$c(HCO_3^-) = x(HCO_3^-) \cdot c(C) \qquad \text{(Eq 6.1-3)}$$

$$c(CO_3^{2-}) = x(CO_3^{2-}) \cdot c(C) \qquad \text{(Eq 6.1-4)}$$

The example calculation was performed for **400 ppmv**. The results are listed as an example in Table 6.1-1. (Calculation example 10.8)

Table 6.1-1: Calculation example for the thought model at 10 °C. Concentrations in mmol/L (400 ppmv CO_2)

		CO_2	HCO_3^-	CO_3^{2-}	C
pH 5.6	Share	0.880	0.120	$0.155 \cdot 10^{-5}$	1
	Concentration	0.0215	0.00293	$0.380 \cdot 10^{-7}$	0.0244
pH 7.0	Share	0.226	0.774	$0.251 \cdot 10^{-3}$	1
	Concentration	0.0215	0.0737	$0.239 \cdot 10^{-4}$	0.0952
pH 8.0	Share	0.0283	0.969	0.00315	1
	Concentration	0.0215	0.737	0.00239	0.761
pH 9.0	Share	0.00282	0.966	0.0314	1
	Concentration	0.0215	7.37	0.239	7.63

The values calculated for pH 5.6 in Table 6.1-1 reflect the conditions, for example, in a rainwater that is in solution equilibrium with the air. The concentration of 0.0215 mmol/L for CO_2 corresponds to the solubility of CO_2. However, 0.00293 mmol/L HCO_3^- has also already formed.

If a higher pH value is present in the water, the **share** of CO_2 in the carbonic acid sum decreases and, while the gas solubility for CO_2 (0.0215 mmol/L) remains constant, more HCO_3^- and finally also CO_3^{2-} are formed in higher concentrations. The carbonic acid sum C (proportion = 1) increases accordingly.

In the graphical representation of the thought model (Figure 6.1-2:)one can see the steep increase of the HCO_3^- and the CO_3^{2-} concentration with increasing pH, whereas the increase of the CO_3^{2-} concentration is steeper (dependence inversely proportional to the square of the H^+ concentration, see also Chap. 4.7 and calculation example 10.6).

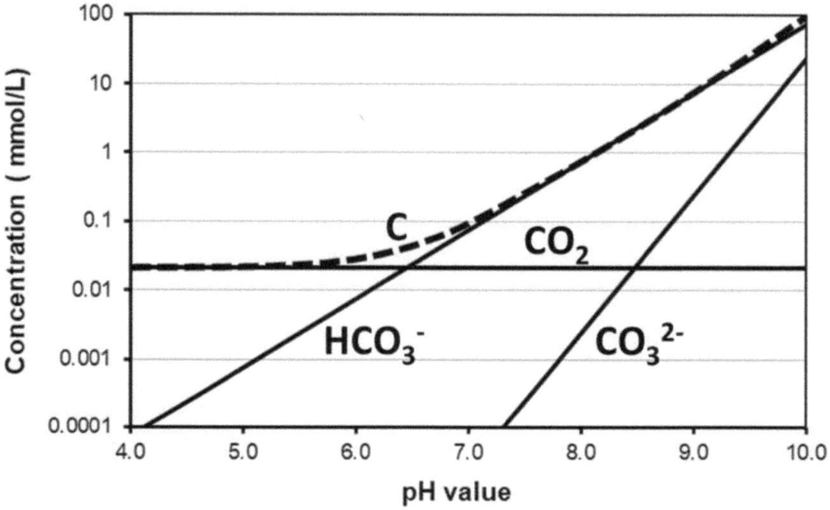

Figure 6.1-2: Semi-logarithmic plot of the concentrations of the carbonic acid forms in an open system as a function of pH (400 ppmv CO_2)

An extended representation of the thought model is shown in Figure 6.1-3 (calculation example 10.8).

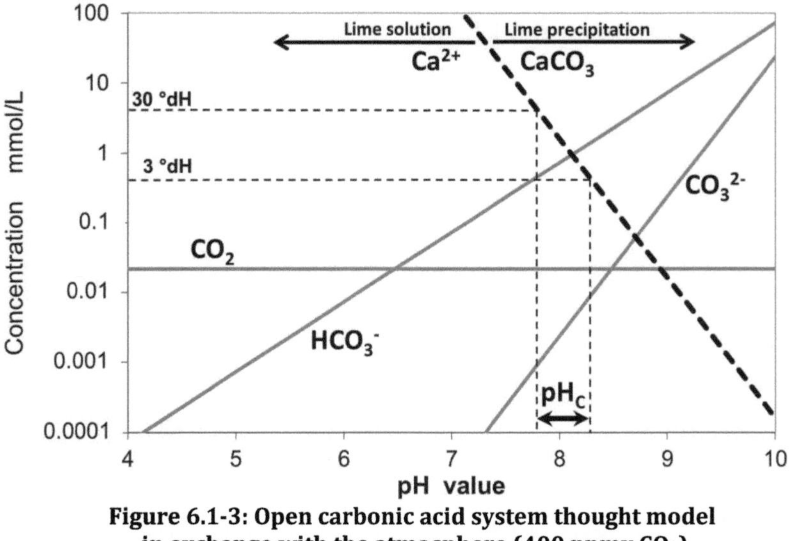

Figure 6.1-3: Open carbonic acid system thought model in exchange with the atmosphere (400 ppmv CO_2)

In addition to the carbonic acid forms, the $Ca^{2+}/CaCO_3$ boundary line of the lime solution or lime precipitation is also drawn here. This line represents for each pH the maximum concentration of free Ca^{2+} as a function of the CO_3^{2-} concentration present in each case (attainment of the solubility product). If this line is exceeded, lime precipitation occurs. If the water falls below the line, it has lime dissolving properties.

Two degrees of hardness are also drawn as guide lines for orientation. Whereby it is assumed that the water hardness is naturally formed to approx. 80% by Ca^{2+}. 5.4 mmol/L (30 °dH) generally form the upper limit for drinking water, 0.5 mmol/L (3 °dH) usually the lower limit (softer drinking waters have to be hardened), so that the section between the two hardness lines represents the theoretical range of drinking waters, but also the generally expected hardness range of surface waters.

The intersection of these hardness lines with the $Ca^{2+}/CaCO_3$ boundary line corresponds to the point where the Ca^{2+} concentration (as a fraction of the hardness) is equal to the solubility product. Projecting these points onto the pH axis, it is clear that the pH of calcite saturation pH_C of waters corresponding to such an **open** system ranges from about pH 7.8 to pH 8.25 (Figure 6.1-3).

In corresponding turbulent streams, which are in lime-carbonic acid equilibrium, pH values of this order are measured (chapter 5.6).

If we transfer this model of thought to the processes during the percolation of water in the subsurface (groundwater recharge), we can assume that the system remains open, i.e. that gas exchange and equilibration with the soil air is possible.

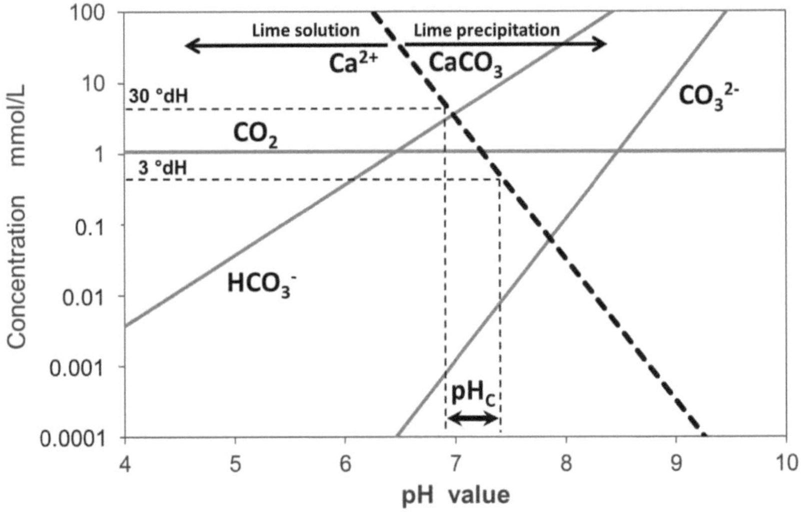

Figure 6.1-4: Changes in the open carbonic acid system due to higher partial pressure in the soil air (2 vol% CO_2)

However, the partial pressure of CO_2 in the soil air is increased (see also chapter 4.1), so that higher concentrations of CO_2 occur in the water. Depending on this, the other carbonic acid components in the thought model also change.

For the presentation in Figure 6.1-4: a CO_2 content of the soil air of 2.0% by volume was used as a basis. Due to the fact that a comparably higher concentration of CO_3^{2-} is present, the limit for lime precipitation is already reached at lower pH values. At the soil gas concentration of 2% by volume used for the calculation, the equilibrium pH_C would be approximately 6.9 - 7.4.

If the individual values for a water with pH 7 (Table 6.1-2:) are considered in this system, the concentrations of the carbonic acid forms can be assigned to the corresponding analysis values (Chapter 7.2).

The CO_2 concentration corresponds to the base capacity 8.2 and the HCO_3^- concentration to the acid capacity 4.3. The carbonic acid sum is recorded via the TIC.

Table 6.1-2: Values of the carbonic acid model for soil air with 2.0% by volume CO_2

pH 7.0		CO_2	HCO_3^-	CO_3^{2-}	C
	Share	0.226	0.774	$0.251 \cdot 10^{-3}$	1
	Concentration	1.076	3.69	0.00120	4.76
		$\approx K_{B8,2}$!	$\approx K_{S4,3}$!		TIC !

These values correspond in magnitude to the values of a groundwater analysis.

6.2 Closed system

If, on the other hand, we consider a **closed system**, the most significant change from the previous consideration is that **no gas exchange** with the environment is possible.

This idea is given, for example, when the seepage water has reached the groundwater horizon during groundwater recharge and continues to flow horizontally enclosed in the total groundwater body. Gas exchange with the soil air is at most given at the groundwater surface. The rest of the groundwater body is closed from gas exchange.

This situation also applies to water extraction for drinking water treatment. The water from the natural water cycle is captured with a certain amount of carbonic acid c(C) corresponding to its origin and a corresponding pH value, i.e. it is taken up and pumped in pipes. Gas exchange is prevented.

Closed system - no gas exchange with the environment possible

$$CO_2 + H_2O \rightleftharpoons HCO_3^- + H^+ \rightleftharpoons CO_3^{2-} + 2\,H^+$$

The characteristic of the closed system is that the carbonic acid sum c(C) remains constant, while the proportions of the carbonic acid forms and thus their concentrations change as a function of pH.

$$\mathbf{c(C)} = c(CO_2) + c(HCO_3^-) + c(CO_3^{2-}) \quad = \mathbf{constant} \qquad \text{(Eq 6.2-1)}$$

Compared to the open system, the CO_2 concentration changes here when the pH value changes, since no equilibrium adjustment is possible here through gas exchange with the environment.

The graphical representation of the ratios in the closed system corresponds to the relative distribution of the carbonic acid forms as shown in chapter 4.5.3 (Figure 4.5-3:).

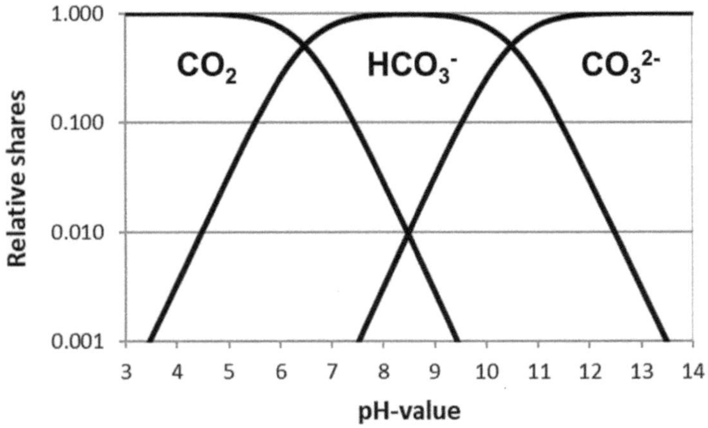

Figure 6.2-1: Figure 4.5-3: Relative proportions of carbonic acid forms as a function of pH (10 °C) (I=0), semi-logarithmic plot

7 Basics for calculations

The carbonic acid system is very complex with high required accuracy. Each user must decide for himself what accuracy and scope he needs.

7.1 Calculation of the constants in the carbonic acid system

For calculations in the carbonic acid system, assigned exact analysis data (chap. 7.2) as well as knowledge of the values for the action constants assigned to the reactions are required as a basis for quantitative statements.

Depending on the accuracy requirement, the effort increases. Often, observations in the basic system of carbonic acid are sufficient (without complex formation).

For accurate calculations, the complex formations and the corresponding complex formation factors must be additionally taken into account. This increases the computational effort considerably.

Basically, one must distinguish whether the calculations are performed for a system open to the atmosphere/gas-phase with possible gas exchange or for a closed system without the possibility of gas exchange. In the open system, in addition to the other equilibrium constants (chap. 7.1.2), the solubility constant (HENRY constant) (chap. 7.1.1) is also required for the exchange of CO_2 with the gas phase.

7.1.1 Solubility constants for CO_2

HENRY's law applies to the calculation of the solubility of CO_2 in water.

$$c_W = K_H \cdot p \qquad \text{(Eq 7.1-1)}$$

$$p = \frac{Vol.\%}{100} \cdot p_{ges} \qquad \text{(Eq 7.1-2)}$$

c_W	Concentration of gas in water [mol/m³]
K_H	HENRY constant [mol/(m³·bar)]
p	Partial pressure [bar]
p_{ges}	Total pressure [bar]

Annotation: The HENRY constant is still used in the literature with the Unit specified in bar (1 bar = 10^5 Pa ; 1 mbar = 1 hPa)

The temperature dependence of the HENRY constant K_H can be calculated by algorithms. One must distinguish between fresh and salt water in the calculation. For fresh water, the ionic strength is generally not taken into account (I=0), but for salt water it must be taken into account.

The solubility constants apply to the **extended CO_2** = $CO_{2(w)}$ + H_2CO_3 (Chap. 4.3).

HENRY constant K_H for freshwater (ionic strength I = 0)

Calculation according to PLUMMER and BUSENBERG [6]:

$$lg(K_H)=108.3865+0.01985076\cdot T-6919.53/T-40.45154\cdot lg(T)+669365/(T)^2 \quad \text{(Eq. 7.1-3)}$$

T Temperature in K ; T = t + 273.15
t Temperature in °C

$$K_H = 10^{lgKH} \qquad [mol/(L\text{-}atm)] \quad \text{(Eq 7.1-4)}$$

Conversion of units:

$$[K_H] = [mol/(L\cdot atm)] \cdot (10^3/(1.01325\ bar/atm)) = [mol/(m^3\cdot bar)] \quad \text{(Eq. 7.1-5)}$$

K_H for salt water (ionic strength I is taken into account)

For the calculation of the HENRY constants for salt water , EDMOND and GIESKES [64] have given an algorithm taking into account the ionic strength. The ionic strength I can be calculated via the content of chloride *Cl*(‰) (chlorinity).

$$-log\ K_H=-2385.73/T+14.0184-0.0152642\cdot T+I(0.28596-6.167\cdot 10^{4}\cdot T) \quad \text{(Eq 7.1-8)}$$

$$I = 0.00147 + 0.03592\cdot Cl + 0.000068\cdot Cl^2 \quad \text{(Eq 7.1-9)}$$

For the relationship between salinity (salinity *S*) and chlorine content (chlorinity *Cl*), according to a UNESCO recommendation:

$$Salinity[‰] = 1.80655 \cdot Chlorinity[‰] \quad \text{(Eq 7.1-10)}$$

The mean salinity of the oceans is given as 35‰, thus the mean chlorinity is 19.37‰.

Table 7.1-1: HENRY constant K_H of CO_2 for fresh water [6] and for salt water (*S*=35.0‰ and *Cl*=19.37‰) as a function of temperature (see also table chap. 12.1)

Temperature °C		0	10	20
K_H [mol/(m³·bar)]	Freshwater	76.895	53.077	38.674
	Salt water	62.303	43.969	32.571

Table 7.1-2: HENRY constant K_H (25°C) as a function of salinity as well as equilibrium concentration of CO_2 in saltwater with different salinities

Marine Area	*S* [‰]	*Cl* [‰]	K_H [mol/(m³·bar)]	CO_2 [mg/L]
Average	35.0	19.37	28.503	0.508
Mediterranean Sea	37.4	20.70	28.161	0.502
Atlantic	35.4	19.60	28.444	0.507
North Sea	35.0	19.37	28.503	0.508
Pacific	34.5	19.10	28.573	0.510
Baltic Sea	8.0	4.43	32.520	0.580

BUNSEN absorption coefficient α

The solution of CO_2 in water can also be calculated using the **BUNSEN absorption coefficient α**. Through it, the distribution between the phases (gas and water) becomes clear. However, calculations must take into account that α is related to the physical standard state (273.15 K and 101.325 Pa)!

BUNSEN absorption coefficient α

$$\frac{c_W}{c_G} = \text{constant} = \alpha \qquad \text{(Eq 7.1-11)}$$

c_W : Molar concentration in water
c_G : Molar concentration in the gas

α is, like HENRY's constant K_H, dependent on temperature!

The BUNSEN absorption coefficient for CO_2

$$\alpha(CO_2) = 1.205 \qquad \text{at } 10\ °C$$

shows that CO_2 is very soluble in water. For comparison, oxygen is poorly soluble with a coefficient of $\alpha(O_2) = 0.038$.

If α is not known directly, it can be calculated from K_H [49].

<u>Conversions:</u>

$$K_H = \frac{\alpha \cdot 10^3}{22.414 \cdot 1.01325} \qquad \alpha\ [1/\text{atm}] \qquad \text{(Eq 7.1-12)}$$

$$K_H = 44.03 \cdot \alpha \qquad [\text{mol}/(\text{m}^3 \cdot \text{bar})] \qquad \text{(Eq 7.1-13)}$$

In this paper, only HENRY's constant is used for the gas solubility calculations.

7.1.2 The equilibrium constants in the base system

For calculations in the base system of carbonic acid (without taking complex formation into account), the dissociation constants K_1 and K_2 as well as the ion product of water K_W and the solubility product K_C of $CaCO_3$ are required. In an open system, the solubility for CO_2 must also be taken into account (see above Chap. 7.1.1)

Dissociation constant 1: $$\frac{a(H^+)\cdot a(HCO_3^-)}{a(CO_2)} = K_1 \quad [(mol/L)] \qquad \text{(Eq 7.1-11)}$$

Dissociation constant 2: $$\frac{a(CO_3^{2-})\cdot a(H^+)}{a(HCO_3^-)} = K_2 \quad [(mol/L)] \qquad \text{(Eq 7.1-12)}$$

Ionic product of the water: $a(H^+)\cdot a(OH^-) = K_W$ $[(mol/L)^2]$ (Eq 7.1-13)

Solubility product ($CaCO_3$): $a(Ca^{2+})\cdot a(CO_3^{2-}) = K_C$ $[(mol/L)^2]$ (Eq 7.1-14)

(For the ion product of water, see chap. 11.4)

The dependence of the constants on temperature is generally represented by the so-called Van't Hoff equation [65]. The basis of the calculations is in each case the constant K_0 (or lgK_0) specific to each reaction at 25 °C, taking into account the temperature dependence of the enthalpy of reaction ΔH and the heat capacity C_p [66] [44] [65].

> <u>Annotation:</u> *The values for lgK_0 belong to the respective reaction equation given. If the direction of the reaction is reversed, lgK_0 changes sign.*

$$\lg(K) = \lg(K_0) + \frac{\Delta H}{\ln(10)\cdot R}\cdot\left(\frac{1}{T_0} - \frac{1}{T}\right) + \frac{C_P}{\ln(10)\cdot R}\cdot\left[\ln\left(\frac{T}{T_0}\right) + \frac{T_0}{T} - 1\right] \qquad \text{(Eq 7.1-15)}$$

By this mean:

lg(K)	Decad. Log. of equilibrium constants at temperature T
$lg(K_0)$	Decad. Log. of equilibrium constants at reference temperature T_0
T	Temperature in K ; T = t[°C] + 273.15
T_0	Reference temperature 298.15 K (25 °C)
ΔH	Reaction enthalpy, J/mol
C_P	Heat capacity of the reaction, J/(mol·K)
R	Universal gas constant 8.314 J/(mol·K)
ln(10)	= 2.302585

For the required constants, values with slight variations can be found in the literature, especially the data for ΔH and Cp vary [63] [65] [66] [67] [68] .

For simplification one can write under introduction of two constants A and B:

$$\lg(K) = \lg(K_0) + A \cdot \left(\frac{1}{T_0} - \frac{1}{T}\right) + B \cdot \left[\ln\left(\frac{T}{T_0}\right) + \frac{T_0}{T} - 1\right] \qquad \text{(Eq 7.1-16)}$$

$$\text{With:} \quad A = \frac{\Delta H}{\ln(10) \cdot R} \qquad \text{(Eq 7.1-17)}$$

$$B = \frac{C_P}{\ln(10) \cdot R} \qquad \text{(Eq 7.1-18)}$$

Table 7.1-2: Constants in the base system [63] [66] [67] [68]

Reaction equations			Formula-characters	$\lg(K_0)$ at 25 °C	A	B
$CO_2 + H_2O$	⇌	$HCO_3^- + H^+$	K_1	-6.356	483.19	-17.238
HCO_3-	⇌	$CO_3^{2-} + H^+$	K_2	-10.329	780.94	-15.149
H_2O	⇌	$H^+ + OH^-$	K_W	-13.996	2953.0	-10.291
$CaCO_3$	⇌	$Ca^{2+} + CO_3^{2-}$	K_C	-8.481	-522.37	-13.059

Annotation: Any deviating signs in literature sources result from the different presentation of the reaction equations.

The following equations with the appropriate constants were used to calculate the temperature dependence.

$\lg(K_1)$= -6.356 + 483.19·[(1/T_0)-(1/T)] -17.238·[ln(T/T_0)+(T_0/T)-1] (Eq 7.1-19)

$\lg(K_2)$= -10.329 + 780.94·[(1/T_0)-(1/T)] -15.149·[ln(T/T_0)+(T_0/T)-1] (Eq 7.1-20)

$\lg(K_W)$=-13.996 + 2953.0·[(1/T_0)-(1/T)] -10.291·[ln(T/T_0)+(T_0/T)-1] (Eq 7.1-21)

$\lg(K_C)$= -8.481 - 522.37·[(1/T_0)-(1/T)] - 13.059·[ln(T/T_0)+(T_0/T)-1] (Eq 7.1-22)

In Table 7.1-3 calculated values for the constants in the base system are shown. Further data for other temperatures are listed in chapter 12.2. The calculated constants correspond to the values calculated according to DIN. For calculations according to DIN 38404-10 (2012)[66] the exact constants specified in DIN should be used to obtain comparable results with the sample calculations.

Table 7.1-3: Values for the constants in the base system

Temperature °C	K_1 mol/L	K_2 mol/L	K_W (mol/L)^2	K_C (mol/L)^2
0	$2.677\cdot10^{-7}$	$2.352\cdot10^{-11}$	$1.140\cdot10^{-15}$	$4.244\cdot10^{-9}$
5	$3.054\cdot10^{-7}$	$2.788\cdot10^{-11}$	$1.847\cdot10^{-15}$	$4.100\cdot10^{-9}$
10	$3.426\cdot10^{-7}$	$3.249\cdot10^{-11}$	$2.920\cdot10^{-15}$	$3.927\cdot10^{-9}$
15	$3.781\cdot10^{-7}$	$3.725\cdot10^{-11}$	$4.510\cdot10^{-15}$	$3.733\cdot10^{-9}$
20	$4.110\cdot10^{-7}$	$4.209\cdot10^{-11}$	$6.817\cdot10^{-15}$	$3.524\cdot10^{-9}$
25	$4.406\cdot10^{-7}$	$4.688\cdot10^{-11}$	$1.009\cdot10^{-14}$	$3.304\cdot10^{-9}$

7.2 Analytical detection of carbonic acid forms

7.2.1 TIC and DIC

TIC = Total Inorganic Carbon

DIC = Dissolved Inorganic Carbon

The direct way to analyze carbonic acid is to record the carbonic acid sum c(C) by determining the TIC.

It is an academic discussion whether to call the parameter DIC or TIC. Formally, if DIC is determined, by definition the sample would have to be filtered through 0.45 µm before analysis to ensure that no particulate carbon is measured. However, filtration is not possible because of the CO_2 , as its concentration could change due to outgassing or input from the air, so that the DIC cannot be detected analytically for technical reasons. Therefore, strictly speaking, the analytical parameter should be called TIC because of the lack of filtration.

The DIN 38404(10) [66] however, uses the designation c(DIC) to denote the carbonic acid sum c(C).

Sampling:

When taking samples for TIC analysis, the general regulations for taking samples for gas analysis must be observed. It is advisable to use 100 ml oxygen bottles with beveled glass stoppers. The sample water is introduced deep into the bottle using a hose. Allowing the bottle to overflow displaces all air from the bottle, so that the sample water in the bottle afterwards had no contact with the air. The bottle is closed without air bubbles with the help of the slanted stoppers.

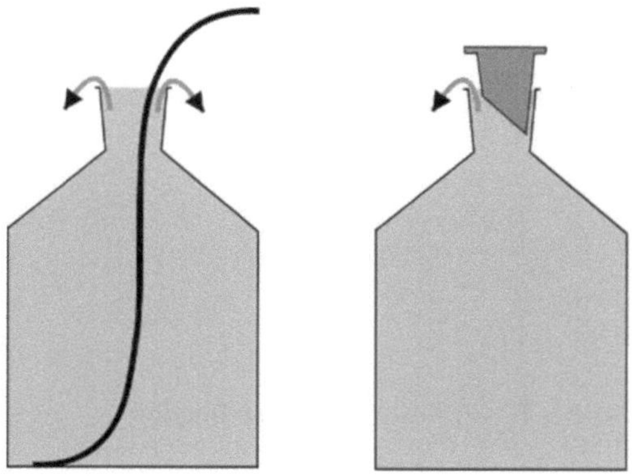

Figure 7.2-1: Sampling bottle with inclined stopper

It is essential that the bottles are transported refrigerated to avoid outgassing due to temperature rise. The analyses should be performed as soon as possible. Self-priming analyzers that take the sample directly from the sampling bottle have proven to be advantageous as there is then no need for further manipulation with the water, which could lead to further errors due to gas exchange. The pH value of the water must be measured directly when taking the sample!

Measuring principle :

Acidifying the sample (usually to pH 2) shifts the equilibrium system of carbonic acid forms completely to the side of CO_2 (Figure 7.2-2:) (Chap. 4.5). This means that all the carbonic acid is present as CO_2. The (gaseous) CO_2 is then expelled using a CO_2-free gas and quantified.

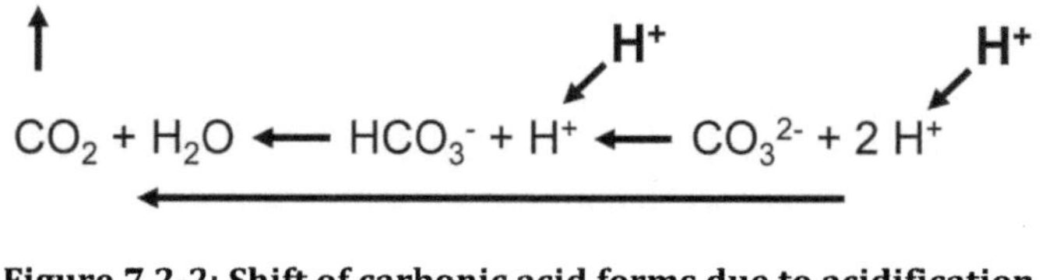

Figure 7.2-2: Shift of carbonic acid forms due to acidification

Usually, CO_2 in the gas phase is measured by absorption in the infrared range. This determination is usually carried out with TOC/TIC measuring equipment, in which case the oxidation state is switched off or bypassed (Figure 7.2-3).

(TOC = Total Organic Carbon)

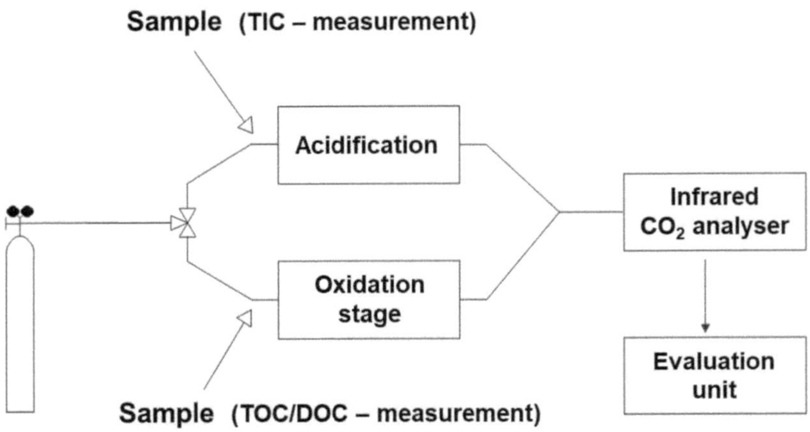

Figure 7.2-3: Schematic diagram of a TOC/TIC measuring apparatus

The molar fractions x(i) of the corresponding forms of carbonic acid are calculated as a function of pH using the equilibrium constants (calculation example 10.7). Multiplying the mass fractions by the carbonic acid sum c(C) (TIC) gives the respective molar concentrations of the carbonic acid forms (calculation example 10.9).

$$c(C) = TIC \quad \rightarrow \quad c(CO_2) = x(CO_2) \cdot c(C) \qquad \text{(Eq 7.2-1)}$$

$$c(HCO_3^-) = x(HCO_3^-) \cdot c(C) \qquad \text{(Eq 7.2-2)}$$

$$c(CO_3^{2-}) = x(CO_3^{2-}) \cdot c(C) \qquad \text{(Eq 7.2-3)}$$

7.2.2 Determination of acid and base capacity

Compared to the direct determination of the TIC, the more common way to determine the carbonic acid sum in practice is the indirect determination by calculation from the acid and base capacity of a water.

When determining the acid and base capacity, however, it must be ensured that no other weak acids or bases (for example humic acids) are present in higher concentrations in the water apart from carbonic acid, which would increase the consumption of acid or base via their buffering behavior (then the TIC must be determined for accurate measurements!).

If the water sample is ultrapure water, only the dissociation equilibrium of the water has to be shifted by titration with H^+ or OH^-. If substances are present in the water which react as weak acids or bases, these substances are detected by the titration, since they react with H^+ or with OH^- and thus lead to a higher consumption during the titration.

Table 7.2-1: shows examples of weak bases which, when present in water, lead to increased H^+ consumption when titrated with acid.

In general, it is assumed that in natural waters, except for carbonic acid, all other weak bases/acids are present only in low concentrations and can be neglected. Under this assumption, the measured value of the acid capacity 4.3 is used as the quantitative determination of the HCO_3^-- and CO_3^{2-}-ions and the measured value of the base capacity 8.2 as the quantitative determination of the CO_2 in the water.

Table 7.2-1: Reactions of weak bases with H^+

Compound	weak Base	Reaction under H^+ consumption to the conjugate acid
Carbon dioxide	HCO_3^-	$+ H^+ \longrightarrow H_2CO_3$
	CO_3^{2-}	$+ H^+ \longrightarrow HCO_3^-$
Phosphate	$H_2PO_4^-$	$+ H^+ \longrightarrow H_3PO_4$
	HPO_4^{2-}	$+ H^+ \longrightarrow H_2PO_4^-$
Ammonia	NH_3	$+ H^+ \longrightarrow NH_4^+$
Boric acid	$B(OH)^{4-}$	$+ H^+ \longrightarrow H_3BO_3 + H_2O$
Organ. Acids	R-COO$^-$	$+ H^+ \longrightarrow$ R-COOH

Notes:

- In seawater, the content of boric acid must be taken into account if m-value and c(C) are to be determined with an accuracy better than 2% [63].
- Ammonium must be considered when titrating in rainwater.
- Phosphate levels must be determined for phosphated drinking waters.
- In waters with low carbonic acid contents, buffering by silica at pH > 9 can play a role, e.g., in biogenic decalcification or in boiler feed waters [63].

7.2.2.1 Acid capacity (Alkalinity)

The hydrogen carbonate ion can react as either a weak acid or a weak base. This property is called amphoteric.

If a water is titrated with acid, the hydrogen carbonate reacts as a weak base with the H^+ ions to form the weak acid CO_2 (Eq. 7.2-4).

This acid capacity (alkalinity, acid binding capacity SBV) of a water tells how much acid (H^+ [mmol/L]) must be added to a water sample during a titration to lower the pH of the sample to a certain value. In water analysis, the acid capacity is generally determined up to pH 4.3. At this pH, the carbonic acid is almost completely present as CO_2. Historically, the pH indicator methyl orange with the turnover point at pH 4.3 was used to indicate the end point of the titration.

By adding acid (H^+ ions), the hydrogen carbonate in the water can be converted into CO_2 (initial pH value below 8.2):

pH < 8.2: $$CO_2 + H_2O \longleftarrow HCO_3^- + H^+ \swarrow H^+$$ (Eq 7.2-4)

Therefore, the water is titrated with acid to pH 4.3 (originally the turnover point of the indicator **M**ethyl orange) and this value is called the

$$\text{Acid capacity up to pH 4.3} = K_{S4.3}$$

This value is also often approximated to the **m**-value (see section 7.2.3).

$$K_{S4.3} \approx m \quad \text{(Eq 7.2-5)}$$

The determination of the acid capacity is to be carried out in contact with the air with strong stirring of the sample so that the CO_2 produced can outgas.

At pH values of the water below 8.2, one essentially records the hydrogen carbonate ions.

pH < 8.2 :

$$n(H^+)\text{titrated} \approx n(HCO_3^-) \quad \text{(Eq 7.2-6)}$$

$$K_{S4.3} \approx c(HCO_3^-) \quad \text{(Eq 7.2-7)}$$

If the pH value of the sample is above 8.2, carbonate ions are then also present in higher concentrations in the water. Two H^+ ions are then required to convert the carbonate ions into CO_2 .

pH > 8,2: $HCO_3^- \longleftarrow CO_3^{2-} + H^+$ (with H^+ added) (Eq 7.2-8)

$CO_2 + H_2O \longleftarrow HCO_3^- + H^+$ (with H^+ added) (Eq 7.2-9)

respectively $CO_2 + H_2O \longleftarrow CO_3^{2-} + \mathbf{2\ H^+}$ (Eq 7.2-10)

Thus, at water pH values above 8.2, one records the hydrogen carbonate ions and the carbonate ions.

pH > 8.2 : $n(H^+)_{titrated} \approx n(HCO_3^-) + \mathbf{2n}(CO_3^{2-})$ (Eq 7.2-11)

It follows : $K_{S4.3} \approx c(HCO_3^-) + 2c(CO_3^{2-})$ (Eq 7.2-12)

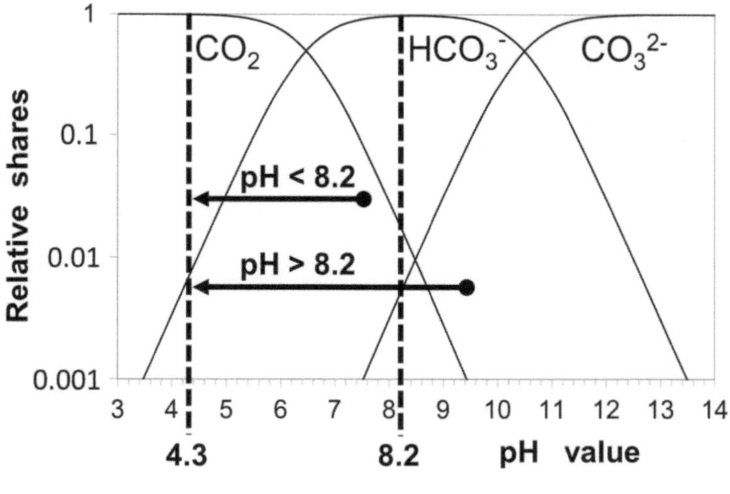

Figure 7.2-4: Representation of the titration possibilities for the acid capacity 4.3

If the pH value of the sample water is above pH 8.2, an acid capacity up to pH 8.2 ($K_{S8.2}$) can also be determined. The titration result corresponds approximately to the concentration of the carbonate ions (Figure 7.2-5:).

$$K_{S8.2} \approx c(CO_3^{2-}) \quad \text{(Eq 7.2-13)}$$

Since this titration is usually too inaccurate in natural waters due to the small titration volume, only the acid capacity 4.3 is determined, even at pH values above 8.2, and hydrogen carbonate and carbonate are recorded together.

7.2.2.2 Base capacity

Analogous to the determination of the acid capacity, the water can be titrated with a base (alkali) up to pH 8.2 (turnover point of the indicator **P**henolphthalein) and called the

Base capacity up to pH 8.2 = $K_{B8.2}$

This value is also often approximated to the **-p-value** (chap. 7.2.3).

$$K_{B8.2} \approx -p \qquad \text{(Eq 7.2-14)}$$

This essentially measures the CO_2 dissolved in the water. The titration converts the gaseously dissolved CO_2 into HCO_3^-.

simplified representation: $CO_2 + OH^- \longrightarrow HCO_3^-$ (Eq 7.2-15)

actual reaction sequence:

$$CO_2 + H_2O \longrightarrow HCO_3^- + H^+ \qquad \mathbf{OH^-} \qquad \text{(Eq 7.2-16)}$$

$$H_2O \longleftarrow H^+ + OH^- \qquad \text{(Eq 7.2-17)}$$

The titration value corresponds approximately to the content of CO_2 in the water:

$$n(OH^-)\text{titrated} \approx n(CO_2) \qquad \text{(Eq 7.2-18)}$$

It follows : $K_{B8.2} \approx c(CO_2)$ (Eq 7.2-19)

When determining the base capacity, contact with the carbon dioxide in the air must be avoided as far as possible, otherwise CO_2 will constantly dissolve from the air during the titration (stir only very carefully and slowly !). Under certain circumstances - especially with soft water - there is a major source of error here. Then it is better and more accurate to calculate the base capacity from acid capacity and pH. (Regarding the delayed reaction speed, see chap. 4.3)

In Figure 7.2-5: the acid and base capacities relevant for the carbonic acid system are shown in an overview.
Considering as an example a water 1 (X) with pH < 8.2, $K_{S4.3}$ and $K_{B8.2}$ could be determined in this water.
For a water 2 (Y) with pH > 8.2, $K_{S4.3}$ and $K_{S8.2}$ would be determinable, although the determination of $K_{S8.2}$ is usually omitted, as shown above.

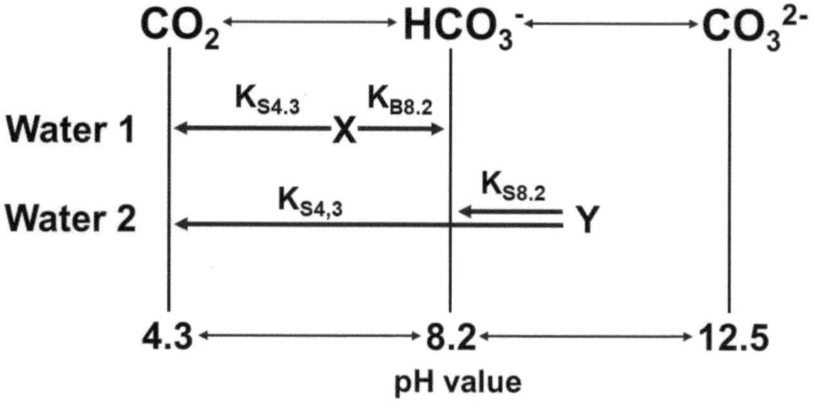

Figure 7.2-5: Overview of the acid and base capacities relevant for the carbonic acid system

7.2.2.3 Practical determination of acid and base capacity

<u>**Scope**</u>

$K_{S4.3}$	only determinable if	pH > 4.3
$K_{B8.2}$	only determinable if	pH < 8.2
$K_{S8.2}$	only determinable if	pH > 8.2

<u>**Sampling**</u>

Sampling for acid capacity can be done normally without any special precautions (see also definition of the m-value chap. 7.2.3.1).

For the determination of the base capacity, a change in the CO_2 concentration must be avoided. The water sample is filled into the glass bottle with a hose. The hose must reach the bottom of the vessel. The sample water must not have any contact with the air! (See also sampling at TIC, chap. 7.2.1)

Allow at least one bottle volume of water to flow over and then close the vessel with a glass stopper to prevent bubbles.

Since outgassing of carbon dioxide is facilitated at elevated temperatures, water samples should be transported in refrigerated bags if possible.

During sampling, the pH and temperature of the water are measured on site.

Determination of $K_{S4.3}$

Determination of acid capacity up to pH 4.3 is performed by titration of the sample (100 ml) with hydrochloric acid (0.1 mol/L) up to pH 4.3.

The carbon dioxide in the air (input/output) has no influence on the determination of the acid capacity! (See also 7.2.3.1)

The temperature is measured at the end of the titration.

Determination of $K_{B8.2}$

Determination of the base capacity up to pH 8.2 is carried out by titration of the sample (100 ml) with sodium hydroxide solution (0.1 mol/L or 0.02 mol/L) up to pH 8.2.
In order to avoid influences of the carbon dioxide of the air (entry/gassing), the water sample must only be stirred slowly.
The adjustment of the pH value during titration is delayed (see also chap. 4.5), therefore it is always necessary to wait a little during the titration steps.
The temperature is measured at the end of the titration.

Indication of the results

The titration results are given in mmol/L rounded to a maximum of 2 decimal places. The temperature measured at the end of the titration must also be indicated.

Ex:	$K_{S4.3}$ = 3.20 mmol/L	at 16 °C
	$K_{B8.2}$ = 0.85 mmol/L	at 15°C
	(Example of a groundwater)	

In Germany, the standard for this determination is DIN 38 409-7: "German standard methods for the examination of water, waste water and sludge - Parameters characterizing effects and substances (group H) - Part 7: Determination of acid and base-neutralizing capacities (H 7)" [69].

7.2.3 m-value and p-value

The definition of the m- and p-values are initially given here without consideration of complexation reactions to facilitate understanding. If high accuracy is required - and only then - the complexation reactions (chap. 7.6) must be taken into account.

m- and p-values are purely **calculated values** and cannot be directly measured analytically. The designation of **m-value** and **p-value** must be understood historically, as these used to be equated with the titration results of the acid capacity 4.3 and base capacity 8.2 of a water, respectively. The initial letters of the indicators used for titration, **M**ethyl orange for $K_{S4.3}$ and **P**henolphthalein for $K_{B8.2}$, led to the naming. The equation of $K_{S4.3}$ and m-value as well as of $K_{B8.2}$ and (-)p-value, which was common in former times, is only possible approximately according to today's definitions, because the respective change of the dissociation of the water by the titration has to be considered.

$$K_{S4.3} \approx m \qquad \text{(Eq 7.2-20)}$$

$$K_{B8.2} \approx -p \qquad \text{(Eq 7.2-21)}$$

The negative sign for p in equation 7.2-21 results from the definition of the p-value (chap. 7.2.3.2).

7.2.3.1 m value

The m-value of a solution is defined by the charge balance of the ions of a water (Eq. 7.2-23 and Eq. 7.2-24). [49] [50]

In an aqueous solution, the positive charges of the cations correspond in their sum to the sum of the negative charges of the anions, so that the solution behaves electrically neutral as a total system (electroneutrality condition). This is represented by a formula in a charge balance.

$$\Sigma \text{ molar charge conc. Cations} = \Sigma \text{ molar charge conc. Anions} \qquad \text{(Eq 7.2-22)}$$

$$\sum_i z(i^+) \cdot c(x_i^+) = \sum_i z(i^-) \cdot c(y_i^-) \qquad \text{(Eq 7.2-23)}$$

z(i) = charge number
x_i^+ = cation
y_i^- = anion

In the charge balance shown here (Eq. 7.2-24), the usual main constituents of a water and the ions of carbonic acid as well as the ions from the dissociation of the water are considered. (In the case of doubly charged ions, the number of charges corresponds to twice the molar concentration.)

Charge Balance:

$$2c(Ca^{2+})+2c(Mg^{2+})+c(Na^{+})+c(K^{+})+c(H^{+})$$
$$=2c(SO_4^{2-})+2c(CO_3^{2-})+c(HCO_3^{-})+c(NO_3^{-})+c(Cl^{-})+c(OH^{-}) \quad \text{(Eq 7.2-24)}$$

Annotation: As a rule, charge balances are also used to check analytical results, since the sums of the concentrations of the charges must be equal if the analytical values are correct. For these practical considerations, $c(CO_3^{2-})$, $c(H^+)$, and $c(OH^-)$ can be omitted because their values are comparatively small.

If the charge balance is specifically rearranged to combine on one side of the equation those ions that resulted from the dissociation of the carbonic acid and water, we obtain :

Rearranged charge balance:

$$2c(Ca^{2+})+2c(Mg^{2+})+c(Na^{+})+c(K^{+})-2c(SO_4^{2-})-c(NO_3^{-})-c(Cl^{-})$$
$$= c(HCO_3^{-})+2c(CO_3^{2-})+c(OH^{-})-c(H^{+}) \quad \text{(Eq 7.2-25)}$$

Ions from the dissociation of the Carbonic acid and of the water !

The sum of the ions from the dissociation of the water and the carbonic acid, i.e., the **right-hand side** of equation (7.2-25), is defined as the m-value. [49] [50]

Definition: $m = c(HCO_3^{-}) + 2c(CO_3^{2-}) + c(OH^{-}) - c(H^{+})$ (Eq 7.2-26)

In this representation the m-value is independent of temperature and ionic strength ! [49]

In an extended definition, one arranges the ions of the weak electrolytes (weak acids and weak bases) on the right side of the charge balance. This definition is useful when, in addition to carbonic acid, other weak electrolytes are present in the water whose concentrations change as a result of titrations and must be taken into account in accurate calculations.

Other weak electrolytes would be, for example, phosphates.

$$m=c(HCO_3^{-})+2c(CO_3^{2-})+c(OH^{-})-c(H^{+})+c(H_2PO_4^{-})+2c(HPO_4^{2-})+3c(PO_4^{3-}) \quad \text{(Eq 7.2-27)}$$

Since in equation (7.2-25) the right side is naturally equal to the left side, it necessarily follows that the m-value is also equal to the left side.

also: $m=2c(Ca^{2+})+2c(Mg^{2+})+c(Na^{+})+c(K^{+})-2c(SO_4^{2-})-c(NO_3^{-})-c(Cl^{-})$ (Eq 7.2-28)

This approach is fundamentally important for the assessment of changes in water composition, for example in water treatment.

Thus, it becomes clear that the addition of a neutral salt, for example sodium chloride, does not change the m-value of the water, since the addition of Na^+ and Cl^- in the equation balance each other out [$+c(Na^+) - c(Cl^-)$].

Also, changes in the CO_2 concentration in the water (addition or withdrawal) do not affect the m-value, since CO_2 is not included in this equation. Considering the right side of the equation (Eq. 7.2-25), this statement would not be so easy to make. (However, this statement is only valid if the change in concentration of CO_2 does not cause precipitation or dissolution of $CaCO_3$)

When assessing the miscibility of waters this m-value can be determined by simple mixing calculation.

For example, if two waters A and B with proportions a and b (a+b=1) are mixed together, the m-value of the mixture can be calculated from the initial m-values.

$$m_{neu} = a \cdot m_A + b \cdot m_B \qquad \text{(Eq 7.2-29)}$$

7.2.3.2 p-value

The m-value and the carbonic acid sum c(C) would be sufficient to describe the carbonic acid system, along with the temperature and ionic strength. Furthermore, in analogy to analytics ($K_{B8.2}$), the p-value has been defined as a link between charge balance and molar balance.

Subtracting the carbonic acid sum c(C) from the m value yields, by definition, the p-value (Eq 7.2-34).

Definition: $p = m - c(C)$ (Eq 7.2-30)

m	=	$c(HCO_3^-) + 2c(CO_3^{2-}) + c(OH^-) - c(H^+)$	(Eq 7.2-31)
(-) $c(C)$	=	$c(CO_2) + c(HCO_3^-) + c(CO_3^{2-})$	(Eq 7.2-32)
$m - c(C) = \mathbf{p}$	=	$- c(CO_2) + c(CO_3^{2-}) + c(OH^-) - c(H^+)$	(Eq 7.2-33)

$$p = - c(CO_2) + c(CO_3^{2-}) + c(OH^-) - c(H^+) \qquad \text{(Eq 7.2-34)}$$

From this definition, the following results for the carbonic acid sum:

$$c(C) = m - p \qquad \text{(Eq 7.2-35)}$$

By this definition, the negative sign in the linkage between p and $K_{B8.2}$ also becomes clear.

$$K_{B8.2} \approx -p \qquad \text{(Eq 7.2-21)}$$

because: $$K_{B8.2} \approx c(CO_2) \qquad \text{(Eq 7.2-19)}$$

and: $$p = -c(CO_2) + c(CO_3^{2-}) + c(OH^-) - c(H^+) \qquad \text{(Eq 7.2-34)}$$

In the presence of other weak electrolytes results accordingly:

$$p=-c(CO_2)+c(CO_3^{2-})+c(OH^-)-c(H^+)+c(H_2PO_4^-)+2c(HPO_4^{2-})+3c(PO_4^{3-}) \qquad \text{(Eq 7.2-36)}$$

7.2.3.3 Calculation of m and p from $K_{S4.3}$ and $K_{B8.2}$

If the condition is met that no other weak acids or bases other than carbonic acid are present in appreciable concentrations in the water, the m-value can be calculated from the acid capacity 4.3 and the p value can be calculated from the base capacity 8.2.

The dilution caused by the titration in the determination of the acid and base capacity is kept very low by selecting a suitable normality of the titration solution so that it does not have to be taken into account.

Calculation of m from $K_{S4.3}$

a) Approximate calculation of the m-value from the acid capacity:

The proportions of HCO_3^- (< 1 %) still present even at pH 4.3 (see chap. 4.5 and Table 7.2-2:) are neglected in the approximate calculation because of the low value.

Table 7.2-2: Relative proportions of the carbonic acid forms Percentages of at pH 4.3 (10 °C) (I=0)

	Relative proportions of the carbonic acid forms		
	$x(CO_2)$	$x(HCO_3^-)$	$x(CO_3^{2-})$
pH 4.3	0.993	0.007	$0.4 \cdot 10^{-8}$

Since the titration of the water with acid consumes a part of the added protons for the change of the pH value, this part of the titrant consumed for the pH change must be taken into account (see also chap. 10.10).

Approximating [49]:

for ionic strengths below $8.3 \cdot 10^{-3}$ mol/L: $m = K_{S4.3} - 5.0 \cdot 10^{-5}$ [mol/L] (Eq. 7.2-37)

for ionic strengths above $8.3 \cdot 10^{-3}$ mol/L: $m = K_{S4.3} - 6.0 \cdot 10^{-5}$ [mol/L] (Eq. 7.2-38)

<u>b) Exact conversion</u>

Since the water titrated to pH 4.3 also has a residual m-value $(\text{m-value})_{4.3}$, this must be added to the titration result ($K_{S4.3}$).

$$m = K_{S4.3} + (\text{m-value})_{4.3} \qquad \text{(Eq 2.2-39)}$$

The m-value at pH 4.3 captures the pH shift considered above and the residual concentration of HCO_3^- at pH 4.3. The carbonate concentration no longer plays a role at pH 4.3 (Table 7.2-2:).

If a **full analysis is available**, the m-value(4.3) can be calculated from the titration result (here titration with HCl) (derived from equation 7.2-28).

$$(\text{m-value})_{4.3} = 2c(Ca) + 2c(Mg) + c(Na) + c(K) - 2c(SO_4) - c(NO_3) - c(Cl) - c(Cl)_{\textbf{titrated}} \qquad \text{(Eq 2.2-40)}$$

<u>**Without full analysis**</u> the m-value(4.3) can be derived from the titration result.

$$K_{S4.3} \approx c(HCO_3^-) \qquad \text{(Eq 7.2-7)}$$

$$\Rightarrow \quad c(C) = \frac{c(HCO_3^-)}{x(HCO_3^-)} \qquad \text{(Eq 7.2-41)}$$

$$\Rightarrow \quad c(C) \approx \frac{K_{S4.3}}{x(HCO_3^-)} \qquad \text{(Eq 7.2-42)}$$

$$(\text{m-value})_{4.3} \approx c(C) \cdot x(HCO_3^-)_{4.3} + 2c(C) \cdot x(CO_3^{2-})_{4.3} + c(OH^-)_{4.3} - c(H^+)_{4.3} \qquad \text{(Eq 7.2-43)}$$

The accuracy of this result can be increased by iteration.

Calculation of p from $K_{B8.2}$

Often the p-value is calculated from m and c(C), since the possibility of error is higher for the determination of the base capacity up to pH 8.2 than for the acid capacity.

For the **exact** calculation of the p-value from the base capacity follows according to the derivation from the m-value:

$$p = -K_{B8.2} + K_{S8.2} + (\text{p-value})_{8.2} \qquad \text{(Eq 2.2-44)}$$

As analysis variables, $K_{B8.2}$ or $K_{S8.2}$ are to be used alternatively. If the pH value is below 8.2, only $K_{B8.2}$ can be determined, if the pH value is above 8.2 , only $K_{S8.2}$ can be determined.

$$(\text{p-value})_{8.2} = c(CO_2)_{8.2} + 2 \cdot c(CO_3^{2-})_{8.2} + c(OH^-)_{8.2} - c(H^+)_{8.2} \qquad \text{(Eq 2.2-45)}$$

An **approximate** conversion of the base capacity into the p-value is not as easy as with the m-value, since the p-value by definition also has to take into account the concentration of CO_3^{2-}, which is not so easy to determine in an approximation.

Therefore, approximately:

$$p \approx -K_{B8.2} \qquad \text{(Eq 2.2-46)}$$

7.2.4 Equivalence factor φ

It has been found convenient for representation and for calculations in the carbonic acid system to define an equivalence factor φ [49] [50].

φ is defined as the proportion of the charge equivalents of the ions HCO_3^- and CO_3^{2-} in the carbonic acid sum c(C).

Definition:

$$\varphi = \frac{c(HCO_3^-) + 2c(CO_3^{2-})}{c(C)} \qquad \text{(Eq 7.2-47)}$$

Since according to chap. 2.1.2:

$$\frac{c(HCO_3^-)}{c(C)} = x(HCO_3^-) \qquad \text{(Eq 7.2-48)}$$

and

$$\frac{c(CO_3^{2-})}{c(C)} = x(CO_3^{2-}) \qquad \text{(Eq 7.2-49)}$$

thus applies: $\varphi = x(HCO_3^-) + 2 \cdot x(CO_3^{2-})$ (Eq 7.2-50)

Introduces a further simplification,

$$\Delta = c(OH^-) - c(H^+) \quad \text{(Eq 7.2-51)}$$

the m-value can be represented in shorthand notation:

$$m = \varphi - c(C) + \Delta \quad \text{(Eq 7.2-52)}$$

7.3 Calculation of the buffering intensity

Buffering intensity is derived from the slope of the titration curve, which reflects the intensity of the pH change due to added acid or base.

The value of the buffering intensity is defined as the differential change of the added base dc(B) or acid dc(S) in relation to the differential change of the pH value dpH:

$$PI = \frac{dc(B)}{dpH} \quad \text{respectively: } PI = -\frac{dc(S)}{dpH} \quad \text{(Eq 7.3-1)}$$

The amount of base/acid added can also be represented by the change in the m-value dm:

$$PI = \frac{dm}{dpH} \quad \text{(Eq 7.3-2)}$$

According to STUMM and MORGAN, the following applies to a diprotic acid like carbonic acid [44]:

$$PI = 2.3 \cdot [c(C) \cdot K_1 \cdot c(H^+) \frac{c(H^+)^2 + 4K_2 \cdot c(H^+) + K_1 \cdot K_2)}{(c(H^+)^2 + K_1 \cdot c(H^+) + K_1 \cdot K_2)^2} + c(H^+) + c(OH^-)] \quad \text{(Eq 7.3-3)}$$

Using the equivalence factor φ (chap. 7.2.4), the following applies [49]:

$$PI = [c(C)(\varphi + 2x(CO_3^{2-}) - \varphi^2) + (c(OH^-) + c(H^+))] \cdot \ln 10 \quad \text{(Eq 7.3-4)}$$

The buffering intensity for waters in the pH range of 4.3 to 8.2 can be calculated using an approximate equation [50] can be calculated:

$$PI = 2.3 \cdot \frac{(K_{S4.3} - 0.05) \cdot K_{B8.2}}{K_{S4.3} + K_{B8.2} - 0.05} \quad [mmol/L] \quad \text{(Eq 7.3-5)}$$

7.4 The importance of temperature

Chemical reactions and their constants depend on temperature. Thus, if one wants to correctly describe and assess a water in terms of its properties, this statement must always refer to a specific state - i.e. to a specific temperature. As a rule, this is the temperature encountered when taking the sample (sampling temperature t_e). However, a temperature other than the sampling temperature may be of interest for evaluating the water with respect to the lime-carbonic acid system. Therefore, it must be possible to perform a calculation for a specific evaluation temperature t_b for which one wishes to perform the evaluation of the water with respect to its behavior towards lime.

Required are the temperatures during sampling, pH measurement and titration of acid and base capacity (Table 7.4-1:).

Table 7.4-1: Temperature specifications to be considered

Designation	Formula symbol
Temp. at sampling	t_e
Temp. at pH measurement	t_{pH}
Temp. for titration $K_{S4,3}/K_{B8,2}$	t_{tit}
Evaluation temperature	t_b

The pH value of the water should be measured directly when the sample is taken at the appropriate temperature to assess the lime-carbonic acid system, without long transport distances and changing sample vessels between sampling and measurement. However, since this is not always the case, the temperature must be taken into account at the time of the pH measurement. Usually, the temperature at the time of sampling then becomes the evaluation temperature because one wants to evaluate the carbonic acid distribution at the original temperature rather than the distribution that exists in the water at the temperature of the pH measurement.

The pH value is strongly dependent on the temperature and decreases with increasing temperature (Figure 7.4-1:). Therefore, a mathematical correction must be made here if the pH value is not measured directly during sampling.

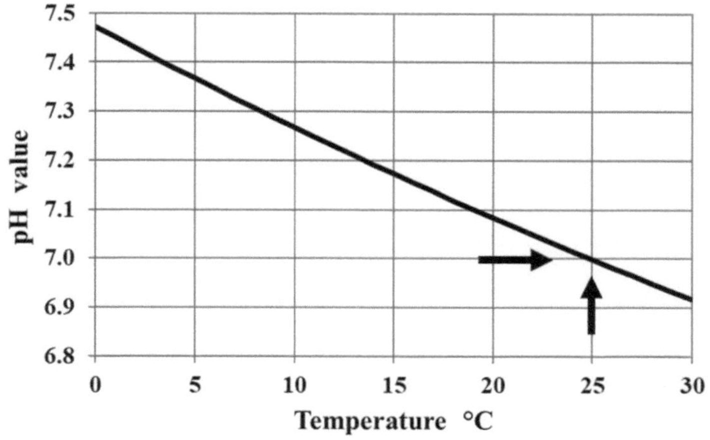

Figure 7.4-1: Temperature dependence of the pH value (pH=7 at 25 °C)

If the temperature (t_{pH}) deviates from the evaluation temperature (t_b) when measuring the pH value, the pH value (pH) measured at this temperature must be converted to the evaluation temperature (pH_{tb}).

$$pH_{tb} = pH + 0.011 \cdot (t_{pH} - t_b) \qquad [52] \qquad \text{(Eq 7.4-1)}$$

The temperature dependence of the conductivity measurement is taken into account analytically by the fact that the measured values are always specified for 25 °C. For the simplified calculation of ionic strength via conductivity (chap. 7.5), the measured value for 25 °C is used.

In the determination of acid capacity, the influence of temperature is small and can usually be neglected [49]. If the base capacity is used as the basis for calculations in the carbonic acid system, the titration temperature must be taken into account and compensated for mathematically.

7.5 Consideration of ionic strength and activity

(Supplementary principles in chapter 11.2)

The ionic strength I of a water is defined as half the sum of the products of the individual ionic concentrations with the squares of the associated charge numbers [49]:

$$I = 0.5 \cdot \sum_{j=1}^{j} [c(i) \cdot (z(i)^2)] \qquad \text{(Eq 7.5-1)}$$

c(i) = concentration of the ion (i) [mol/L]
z(i) = charge number
j = number of ions

The ionic strength is usually calculated using the data from a full analysis with Eq. 7.5-1 calculated.

An approximate calculation of the ionic strength is possible with the measured conductivity (κ_{25}, 25°C) [70]:

$$I = \frac{\kappa_{25}}{6200} \, [\text{mol/L}] \; (\kappa_{25} \text{ in mS/m}) \qquad \text{(Eq 7.5-2)}$$

Note the specified unit of conductivity. With the unit µS/cm, which is also frequently used, the conversion factor changes to 62.000.

An approximate calculation from the evaporation residue R can also be done [71]. Condition: Evaporation residue R < 0.5 g/L.

$$I = 2.5 \cdot 10^{-2} \cdot R \, [\text{mol/L}] \qquad \text{(Eq 7.5-3)}$$

Activity coefficient

The activity coefficient f(i) is the connection between the molar concentration c(i) and the activity a(i):

$$a(i) = c(i) \cdot f(i) \qquad \text{(Eq 7.5-4)}$$

The calculation of the activity coefficient is carried out as a function of the ionic strength I according to DEBEYE-HÜCKEL [49] [66]:

$$\lg(f(i)) = -A \cdot (z(i))^2 \cdot \frac{\sqrt{I}}{1 + B \cdot g(i) \cdot \sqrt{I}} \qquad I \text{ in mol/L} \qquad \text{(Eq 7.5-5)}$$

I = ionic strength of the water
z(i) = charge number
g(i) = Ionsize parameter

As an approximation one can use for the factors [52]:

$$A = 0.5 \qquad B \cdot g(i) = 1.4 \qquad \text{(Eq 7.5-6)}$$

For accurate calculations, one must consider the following relationships [66].

$$A = \frac{1.823 \cdot 10^6}{(T \cdot DK]^{1.5}} \qquad \text{(Eq 7.5-7)}$$

$$B = \frac{50.248}{(T \cdot DK]^{0.5}} \qquad \text{(Eq 7.5-8)}$$

$$DK = 87.74 - 400.08 \cdot 10^{-3} \cdot t + 0.9398 \cdot 10^{-3} \cdot t^2 - 0.00141 \cdot 10^{-3} \cdot t^3 \qquad \text{(Eq 7.5-9)}$$

DK = Dielectric constant of water
T = Temperature in Kelvin
t = Temperature in °C

The value for the corresponding ion size parameters can be obtained from tables in the literature [67] [66].

Using the factors in Equation 7.5-6, the following observations and calculations result.

$$\lg(f(i)) = -0.5 \cdot (z(i))^2 \cdot \frac{\sqrt{I}}{1+1.4 \cdot \sqrt{I}} \qquad \text{(Eq 7.5-10)}$$

For **monovalent** ions $\{(z(i))^2 = 1\}$ then holds:

monovalent ions: $$\lg(f) = -0.5 \cdot \frac{\sqrt{I}}{1+1.4 \cdot \sqrt{I}} \qquad \text{(Eq 7.5-11)}$$

For **divalent** (f_2) and **trivalent** (f_3) ions $(z(i))^2 = 4$ and 9 is derived accordingly:

divalent ions: $$\lg(f_2) = -0.5 \cdot (2)^2 \cdot \frac{\sqrt{I}}{1+1.4 \cdot \sqrt{I}} \qquad \text{(Eq 7.5-12)}$$

$$\lg(f_2) = 4 - \lg(f) \qquad \text{(Eq 7.5-13)}$$

trivalent ions: $$\lg(f_3) = -0.5 \cdot (3)^2 \cdot \frac{\sqrt{I}}{1+1.4 \cdot \sqrt{I}} \qquad \text{(Eq 7.5-14)}$$

$$\lg(f_3) = 9 - \lg(f) \qquad \text{(Eq 7.5-15)}$$

synonymous with:

divalent ions: $$f_2 = f^4 \qquad \text{(Eq 7.5-16)}$$

trivalent ions: $$f_3 = f^9 \qquad \text{(Eq 7.5-17)}$$

eg.: $$a(HCO_3^-) = c(HCO_3^-) \cdot f \qquad \text{(Eq 7.5-18)}$$

$$a(CO_3^{2-}) = c(CO_3^{2-}) \cdot f^4 \qquad \text{(Eq 7.5-19)}$$

To illustrate the influence of ionic strength, the ionic strengths and activity coefficients are shown in Table 7.5-1. The calculation was made for the ionic strength using the electrical conductivity and for the activity coefficients using the factors from Equation 7.5-6.

Table 7.5-1: Activity coefficients as a function of ionic strength

Conductivity mS/m	Ionic strength I mol/L	f_1	$f_2 = f^4$	$f_3 = f^9$
20	0.0032	0.94	0.78	0.58
40	0.0065	0.92	0.72	0.47
60	0.0097	0.91	0.67	0.41
80	0.0129	0.89	0.64	0.36
100	0.0161	0.88	0.61	0.33

For compounds without charges, the activity a(x) is equal to the molar concentration c(x). The associated activity coefficient is then equal to 1.

$$a(x) = c(x) \quad \longrightarrow \quad f(x) = 1 \qquad \text{(Eq 7.5-20)}$$

For the equilibrium constants we get:

$$K = K^* - (f)^{Sz^2} \qquad \text{(Eq 7.5-21)}$$

respectively

$$\lg(K) = \lg(K^*) + Sz^2 \cdot \lg(f) \qquad \text{(Eq 7.5-22)}$$

Sz^2 Sum of the charge squares of the ions involved in the reaction
K Equilibrium constant, dependent only on temperature and pressure {K(T,p)}
K* Equilibrium constant, depending on temperature, pressure and ionic strength {K(T,p,I)}

Illustration using the example of the second dissociation constant of carbonic acid:

$$HCO_3^- \rightleftharpoons CO_3^{2-} + H^+ \qquad \text{(Eq 7.5-23)}$$

$$\frac{a(CO_3^{2-}) \cdot a(H^+)}{a(HCO_3^-)} = K_2 \qquad \text{(Eq 7.5-24)}$$

$$\frac{c(CO_3^{2-}) \cdot c(H^+)}{c(HCO_3^-)} = K_2^* \qquad \text{(Eq 7.5-25)}$$

The activities in Eq 7.5-24 are replaced by the corresponding concentrations and activity coefficients.

$$a(H^+) = c(H^+) \cdot f \qquad \text{(Eq 7.5-26)}$$

$$a(HCO_3^-) = c(HCO_3^-) \cdot f \qquad \text{(Eq 7.5-27)}$$

$$a(CO_3^{2-}) = c(CO_3^{2-}) \cdot f^4 \qquad \text{(Eq 7.5-28)}$$

$$\frac{c(CO_3^{2-}) \cdot f^4 \cdot c(H^+) \cdot f}{c(HCO_3^-) \cdot f} = K_2 \quad \frac{c(CO_3^{2-}) \cdot c(H^+)}{c(HCO_3^-)} \cdot f^4 = K_2 \qquad \text{(Eq 7.5-29)}$$

It follows: $K_2^* \cdot f^4 = K_2$ or $K_2^* = K_2/f^4$ (Eq 7.5-30)

Attention !

For calculations in the carbonic acid system, the concentrations of the substances are determined analytically and inserted accordingly into the formulas, taking into account the activity coefficients. However, since the **activity of** the H^+ ions is recorded directly with the pH electrode, $a(H^+)$ can be derived directly from the pH value and does not have to be taken into account when calculating the activity coefficient!

$$a(H^+) = 10^{-pH} \text{ [mol/L]} \qquad \text{(Eq 7.5-31)}$$

So that for these practical calculations:

$$\frac{c(CO_3^{2-}) \cdot f^4 \cdot a(H^+)}{c(HCO_3^-) \cdot f} = K_2 \qquad \text{(Eq 7.5-32)}$$

It follows: $K_2^* \cdot f^3 = K_2$ (with **a**(H^+) !) (Eq 7.5-33)

For the other mainly used constants then applies accordingly:

$$K_1^* \cdot f = K_1 \qquad \text{(Eq 7.5-34)}$$

$$K_C^* \cdot f^8 = K_C \qquad \text{(Eq 7.5-35)}$$

7.6 Complexation reactions

Beyond the equilibrium reactions presented so far in the general principles, the formation of ion complexes occurs in small proportions [63] [44]. Calcium and magnesium ions as well as hydroxide, hydrogen carbonate, carbonate, sulfate and phosphate ions are involved in these complexation reactions relevant to the carbonic acid system. Sodium, potassium, chloride, and nitrate ions are considered fully dissociated and are not involved in complex formation.

Table 7.6-1: Complexes of calcium and magnesium relevant for the carbonic acid system

Ca complexes	Mg complexes
$CaHCO_3^+$	$MgHCO_3^+$
$CaCO_3^0$	$MgCO_3^0$
$Ca(OH)^+$	$Mg(OH)^+$
$CaH_2PO_4^+$	$MgH_2PO_4^+$
$CaHPO_4$	$MgHPO_4$
$CaPO_4^-$	$MgPO_4^-$
$CaSO_4^0$	$MgSO_4^0$

The complexes are present in the water either as charged complex ions (e.g. $CaHCO_3^+$) or as neutral ion pairs (e.g. $CaCO_3^0$) and affect the reaction equilibria in the carbonic acid system described so far. Therefore, these reactions must be taken into account in accurate calculations. According to an estimation by HÖLZEL [72], a non-consideration leads to errors between about 0.02 and 0.05 pH units, depending on the composition of the water. However, the complexation reactions are not essential for a basic understanding of the carbonic acid system.

The corresponding mass action equations yield the associated mass action constants. The calculation of the constants as a function of temperature is performed according to the Van't Hoff equation (according to chap. 7.1.2). Numerical values for the constants are listed in the appendix (chap. 12.4 and 12.5).

Annotation: The numbering of the constants K_3 to K_{16} in Table 7.6-2: and Table 7.6-3 are only related to this paper.

Table 7.6-2: Complex formations of calcium with constants according to chap. 7.1.2 [63] [66] [67] [68]

				$lg(K_0)$	A	B	
$Ca^{2+} + HCO_3^-$	⇌	$CaHCO_3^+$	K_3	1.121	415.3	0	(Eq 7.6-1)
$Ca^{2+} + CO_3^{2-}$	⇌	$CaCO_3^0$	K_4	3.200	835.8	22.46	(Eq 7.6-2)
$Ca^{2+} + OH^-$	⇌	$Ca(OH)^+$	K_5	1.340	679.1	2.612	(Eq 7.6-3)
$Ca^{2+} + H_2PO_4^-$	⇌	$CaH_2PO_4^+$	K_6	1.006	681.7	-39.70	(Eq 7.6-4)
$Ca^{2+} + HPO_4^{2-}$	⇌	$CaHPO_4$	K_7	2.572	646.2	-26.64	(Eq 7.6-5)
$Ca^{2+} + PO_4^{3-}$	⇌	$CaPO_4^-$	K_8	6.462	600.7	0	(Eq 7.6-6)
$Ca^{2+} + SO_4^{2-}$	⇌	$CaSO_4^0$	K_9	2.310	397.0	8.880	(Eq 7.6-7)

Table 7.6-3: Complex formations of magnesium with constants according to chap. 7.1.2 [63] [66] [67] [68]

				$lg(K_0)$	A	B	
$Mg^{2+} + HCO_3^-$	⇌	$MgHCO_3^+$	K_{10}	1.068	378.7	0	(Eq 7.6-8)
$Mg^{2+} + CO_3^{2-}$	⇌	$MgCO_3^0$	K_{11}	2.947	679.1	3.657	(Eq 7.6-9)
$Mg^{2+} + OH^-$	⇌	$Mg(OH)^+$	K_{12}	2.220	365.7	5.224	(Eq 7.6-10)
$Mg^{2+} + H_2PO_4^-$	⇌	$MgH_2PO_4^+$	K_{13}	1.521	744.4	-38.66	(Eq 7.6-11)
$Mg^{2+} + HPO_4^{2-}$	⇌	$MgHPO_4$	K_{14}	2.878	724.5	-26.64	(Eq 7.6-12)
$Mg^{2+} + PO_4^{3-}$	⇌	$MgPO_4^-$	K_{15}	6.589	679.1	0	(Eq 7.6-13)
$Mg^{2+} + SO_4^{2-}$	⇌	$MgSO_4^0$	K_{16}	2.263	940.3	0.5224	(Eq 7.6-14)

By considering complexation, the basic system of carbonic acid (Section 4.3) is extended (Table 7.6-4:).

Table 7.6-4: Forms of carbonic acid with consideration of complex formation

Carbonic acid - forms with complexes :		
Basic system	CO_2	carbon dioxide dissolved in gas
(simplified system)	HCO_3^-	Hydrogen carbonate ion
	CO_3^{2-}	Carbonate ion
system extended by ion complexes	$CaHCO_3^+$	
	$CaCO_3^0$	
	$MgHCO_3^+$	
	$MgCO_3^0$	

The extension of the carbonic acid system to include the complexation reactions leads to an extended molar balance for the carbonic acid sum. Considering the ion complexes, it is useful to consider the total concentrations c(**ti**) of the species.

Extended molar balance:

$$\mathbf{c(tC)} = c(CO_2) + c(\mathbf{t}HCO_3^-) + c(\mathbf{t}CO_3^{2-}) \qquad \text{(Eq 7.6-15)}$$

with: $$c(tHCO_3^-) = c(HCO_3^-) + c(CaHCO_3^+) + c(MgHCO_3^+) \qquad \text{(Eq 7.6-16)}$$

$$c(tCO_3^{2-}) = c(CO_3^{2-}) + c(CaCO_3^0) + c(MgCO_3^0) \qquad \text{(Eq 7.6-17)}$$

Complexation factors F_{C1} and F_{C2}

Complexation factors F_{C1} and F_{C2} are introduced for simplification in calculations and in the presentation of reaction equations.

$$F_{C1} = \frac{c(tHCO_3^-)}{c(HCO_3^-)} = \frac{c(HCO_3^-)+c(CaHCO_3^+)+c(MgHCO_3^+)}{c(HCO_3^-)} \qquad \text{(Eq 7.6-18)}$$

$$c(tHCO_3^-) = c(HCO_3^-)\cdot F_{C1}$$

$$F_{C2} = \frac{c(tCO_3^{2-})}{c(CO_3^{2-})} = \frac{c(CO_3^{2-})+c(CaCO_3^0)+c(MgCO_3^0)}{c(CO_3^{2-})} \qquad \text{(Eq 7.6-19)}$$

$$c(tCO_3^{2-}) = c(CO_3^{2-})\cdot F_{C2}$$

This gives c(tC) for the total equation of the carbonic acid sum:

$$c(tC) = c(CO_2) + c(HCO_3^-)\cdot F_{C1} + c(CO_3^{2})\cdot F_{C2} \qquad \text{(Eq 7.6-20)}$$

Calculation of F_{C1} and F_{C2} as a function of Ca and Mg concentration

Transformations yield the following equations for the complexation factors F_{C1} and F_{C2} for calculation as a function of Ca and Mg concentration.

$$F_{C1} = \frac{c(HCO_3^-)+c(CaHCO_3^+)+c(MgHCO_3^+)}{c(HCO_3^-)} \qquad \text{(Eq 7.6-18)}$$

$Ca^{2+} + HCO_3^- \rightleftharpoons CaHCO_3^+$

$$\frac{a(CaHCO_3^+)}{a(Ca^{2+})\cdot a(HCO_3^-)} = K_3 \quad \rightarrow \quad \frac{c(CaHCO_3^+)\cdot f}{c(Ca^{2+})\cdot f^4\cdot c(HCO_3^-)\cdot f} = K_3$$

$$\Rightarrow c(CaHCO_3^+) = c(Ca^{2+}) \cdot c(HCO_3^-) \cdot f^4 \cdot K_3$$

$Mg^{2+} + HCO_3^- \rightleftharpoons MgHCO_3^+$

$$\frac{a(MgHCO_3^+)}{a(Mg^{2+})\cdot a(HCO_3^-)} = K_{10} \quad \rightarrow \quad \frac{c(MgHCO_3^+)\cdot f}{c(Mg^{2+})\cdot f^4\cdot c(HCO_3^-)\cdot f} = K_{10}$$

$$\Rightarrow c(MgHCO_3^+) = c(Mg^{2+}) \cdot c(HCO_3^-) \cdot f^4 \cdot K_{10}$$

$$F_{C1} = \frac{c(HCO_3^-)+c(Ca^{2+})\cdot c(HCO_3^-)\cdot f^4\cdot K_3+c(Mg^{2+})\cdot c(HCO_3^-)\cdot f^4\cdot K_{10}}{c(HCO_3^-)}$$

$$\Rightarrow F_{C1} = 1 + c(Ca^{2+}) \cdot f^4 \cdot K_3 + c(Mg^{2+}) \cdot f^4 \cdot K_{10} \qquad \text{(Eq 7.6-20)}$$

$$F_{C2} = \frac{c(CO_3^{2-})+c(CaCO_3^0)+c(MgCO_3^0)}{c(CO_3^{2-})} \qquad \text{(Eq 7.6-19)}$$

$Ca^{2+} + CO_3^{2-} \rightleftharpoons CaCO_3^0$

$$\frac{a(CaCO_3^0)}{a(Ca^{2+})\cdot a(CO_3^{2-})} = K_4 \quad \rightarrow \quad \frac{c(CaCO_3^0)}{c(Ca^{2+})\cdot f^4\cdot c(CO_3^{2-})\cdot f^4} = K_4$$

$$\Rightarrow c(CaCO_3^0) = c(Ca^{2+}) \cdot c(CO_3^{2-}) \cdot f^8 \cdot K_4$$

$Mg^{2+} + CO_3^{2-} \rightleftharpoons MgCO_3^0$

$$\frac{a(MgCO_3^0)}{a(Mg^{2+})\cdot a(CO_3^{2-})} = K_{11} \quad \rightarrow \quad \frac{c(MgCO_3^0)}{c(Mg^{2+})\cdot f^4\cdot c(CO_3^{2-})\cdot f^4} = K_{11}$$

$$\Rightarrow c(MgCO_3^0) = c(Mg^{2+}) \cdot c(CO_3^{2-}) \cdot f^8 \cdot K_{11}$$

$$F_{C2} = \frac{c(CO_3^{2-})+c(Ca^{2+})\cdot c(CO_3^{2-})\cdot f^8\cdot K_4+c(Mg^{2+})\cdot c(CO_3^{2-})\cdot f^8\cdot K_{11}}{c(CO_3^{2-})}$$

$$\Rightarrow F_{C2} = 1 + c(Ca^{2+}) \cdot f^8 \cdot K_4 + c(Mg^{2+}) \cdot f^8 \cdot K_{11} \qquad \text{(Eq 7.6-21)}$$

Other useful complexing factors

Beyond the complexes of the carbonic acid forms, the complexes of Ca and Mg with hydroxide ions and sulfate are also important. This is because it affects the calcium and magnesium concentrations in the other reactions.

$$c(tCa) = c(Ca^{2+}) + c(CaHCO_3^{+}) + c(CaCO_3^{0}) + c(CaOH^{+}) + c(CaSO_4^{0}) \quad \text{(Eq. 7.6-22)}$$

$$c(tMg) = c(Mg^{2+}) + c(MgHCO_3^{+}) + c(MgCO_3^{0}) + c(MgOH^{+}) + c(MgSO_4^{0}) \quad \text{(Eq. 7.6-23)}$$

$$c(tSO_4) = c(SO_4^{2-}) + c(CaSO_4^{0}) + c(MgSO_4^{0}) \quad \text{(Eq 7.6-24)}$$

$$c(tOH) = c(OH^{-}) + c(CaOH^{+}) + c(MgOH^{+}) \quad \text{(Eq 7.6-25)}$$

$$F_{Ca} = \frac{c(tCa^{2+})}{c(Ca^{2+})} = \frac{c(Ca^{2+})+c(CaOH^{+})+c(CaHCO_3^{+})+c(CaCO_3^{0})+c(CaSO_4^{0})}{c(Ca^{2+})} \quad \text{(Eq 7.6-26)}$$

$$Ca^{2+} + HCO_3^{-} \rightleftharpoons CaHCO_3^{+} \qquad \frac{c(CaHCO_3^{+})\cdot f}{c(Ca^{2+})\cdot f^4\cdot c(HCO_3^{-})\cdot f} = K_3$$

$$c(CaHCO_3^{+}) = c(Ca^{2+})\cdot c(HCO_3^{-})\cdot f^4\cdot K_3$$

$$Ca^{2+} + CO_3^{2-} \rightleftharpoons CaCO_3^{0} \qquad \frac{c(CaCO_3^{0})}{c(Ca^{2+})\cdot f^4\cdot c(CO_3^{2-})\cdot f^4} = K_4$$

$$c(CaCO_3^{0}) = c(Ca^{2+})\cdot c(CO_3^{2-})\cdot f^8\cdot K_4$$

$$Ca^{2+} + OH^{-} \rightleftharpoons Ca(OH)^{+} \qquad \frac{c(CaOH^{+})\cdot f}{c(Ca^{2+})\cdot f^4\cdot \mathbf{a}(OH^{-})} = K_5$$

$$c(CaOH^{+}) = c(Ca^{2+})\cdot \mathbf{a}(OH^{-})\cdot f^3\cdot K_5$$

$$Ca^{2+} + SO_4^{2-} \rightleftharpoons CaSO_4^{0} \qquad \frac{c(CaSO_4^{0})}{c(Ca^{2+})\cdot f^4\cdot c(SO_4^{2-})\cdot f^4} = K_9$$

$$c(CaSO_4^{0}) = c(Ca^{2+})\cdot c(SO_4^{2-})\cdot f^8\cdot K_9$$

$$F_{Ca} = \frac{c(Ca^{2+}) + c(Ca^{2+})\cdot c(HCO_3^{-})\cdot f^4\cdot K_3 + c(Ca^{2+})\cdot c(CO_3^{2-})\cdot f^8\cdot K_4}{c(Ca^{2+})} + \frac{c(Ca^{2+})\cdot \mathbf{a}(OH^{-})\cdot f^3\cdot K_5 + c(Ca^{2+})\cdot c(SO_4^{2-})\cdot f^8\cdot K_9}{c(Ca^{2+})}$$

$$F_{Ca} = 1 + c(HCO_3^{-})\cdot f^4\cdot K_3 + c(CO_3^{2-})\cdot f^8\cdot K_4 + \mathbf{a}(OH^{-})\cdot f^3\cdot K_5 + c(SO_4^{2-})\cdot f^8\cdot K_9 \quad \text{(Eq 7.6-27)}$$

$$c(tCa) = c(Ca^{2+})\cdot F_{Ca} \quad \text{(Eq 7.6-28)}$$

With analog derivation follows:

$$F_{Mg} = \frac{c(tMg^{2+})}{c(Mg^{2+})} = \frac{c(Mg^{2+})+c(MgHCO_3^+)+c(MgCO_3^0)+c(MgOH^+)+c(MgSO_4^0)}{c(Mg^{2+})} \qquad \text{(Eq 7.6-29)}$$

$$F_{Mg} = 1 + c(HCO_3^-)\cdot f^4\cdot K_{10} + c(CO_3^{2-})\cdot f^8\cdot K_{11} + \mathbf{a}(OH^-)\cdot f^3\cdot K_{12} + c(SO_4^{2-})\cdot f^8\cdot K_{16} \qquad \text{(Eq 7.6-30)}$$

$$c(tMg) = c(Mg^{2+})\cdot F_{Mg} \qquad \text{(Eq 7.6-31)}$$

$$F_{SO4} = \frac{c(tSO_4^{2-})}{c(SO_4^{2-})} = \frac{c(SO_4^{2-})+c(CaSO_4^0)+c(MgSO_4^0)}{c(SO_4^{2-})} \qquad \text{(Eq 7.6-32)}$$

$$F_{SO4} = 1 + c(Ca^{2+})\cdot f^8\cdot K_9 + c(Mg^{2+})\cdot f^8\cdot K_{16} \qquad \text{(Eq 7.6-33)}$$

$$c(tSO_4) = c(SO_4^{2-})\cdot F_{SO4} \qquad \text{(Eq 7.6-34)}$$

$$F_{OH} = \frac{c(tOH^-)}{c(OH^-)} = \frac{c(OH^-)+c(CaOH^+)+c(MgOH^+)}{c(OH^-)} \qquad \text{(Eq 7.6-35)}$$

$$F_{OH} = 1 + c(Ca^{2+})\cdot f^4\cdot K_5 + c(Mg^{2+})\cdot f^4\cdot K_{12} \qquad \text{(Eq 7.6-36)}$$

$$c(tOH) = c(OH^-)\cdot F_{OH} \qquad \text{(Eq 7.6-37)}$$

7.7 Consideration of the m-value with complex formation

Also for the m-value complex formation must be taken into account for accurate calculations. By introducing the total concentrations c(ti), we obtain accordingly:

$$m=c(tHCO_3^-)+2c(tCO_3^{2-})+c(tOH^-)-c(H^+) \quad \text{(Eq 7.7-1)}$$

$$c(tHCO_3^-)= c(HCO_3^-)+c(CaHCO_3^+)+c(MgHCO_3^+) \quad \text{(Eq 7.7-2)}$$

$$c(tCO_3^{2-}) = c(CO_3^{2-})+c(CaCO_3^0)+c(MgCO_3^0) \quad \text{(Eq 7.7-3)}$$

$$c(tOH^-) = c(OH^-) + c(Ca(OH)^+) + c(Mg(OH)^+) \quad \text{(Eq 7.7-4)}$$

In notation with the complexing factors applies:

$$m = c(HCO_3^-)\cdot F_{C1} +2c(CO_3^{2-})\cdot F_{C2} + c(OH^-)\cdot F_{OH} - c(H^+) \quad \text{(Eq 7.7-5)}$$

Accordingly, also applies:

$$(\text{m value})_{4,3}=c(HCO_3^-)_{4,3}\cdot F_{C1}+2c(CO_3^{2-})_{4,3}\cdot F_{C2}+c(OH^-)_{4,3}\cdot F_{OH}-c(H^+)_{4,3} \quad \text{(Eq 7.7-6)}$$

$$(\text{p value}) = c(CO_2) + 2\cdot c(CO_3^{2-})\cdot F_{C2} + c(OH^-)\cdot F_{OH} - c(H^+) \quad \text{(Eq 7.7-7)}$$

$$(\text{p value})_{8,2} = c(CO_2)_{8,2} + 2\cdot c(CO_3^{2-})_{8,2}\cdot F_{C2} + c(OH^-)_{8,2}\cdot F_{OH}-c(H^+)_{8,2} \quad \text{(Eq 7.7-8)}$$

For the m-value, calculated via the left side of the charge balance (chap. 7.2.3), the complex formation must then also be taken into account.

Again, it makes sense to introduce total concentrations c(ti), since the total concentrations of the parameters in the water do not change due to complex formation. For Na, K, NO_3 and Cl, the simple molar concentration can be used, since these do not form significant complexes.

$$m=2c(tCa)+2c(tMg)+c(Na)+c(K)-2c(tSO_4)-c(NO_3)-c(Cl) \quad \text{(Eq 7.7-9)}$$

$$m=2c(Ca)\cdot F_{Ca}+2c(Mg)\cdot F_{Mg}+c(Na)+c(K)-2c(SO_4)\cdot F_{SO4}-c(NO_3)-c(Cl) \quad \text{(Eq 7.7-10)}$$

In the presence of phosphate, corresponding complexing factors are added:

$$F_{H2PO4} = 1 + c(Ca^{2+})\cdot f^4\cdot K_6 + c(Mg^{2+})\cdot f^4\cdot K_{13} \quad \text{(Eq 7.7-11)}$$

$$F_{HPO4} = 1 + c(Ca^{2+})\cdot f^8\cdot K_7 + c(Mg^{2+})\cdot f^8\cdot K_{14} \quad \text{(Eq 7.7-12)}$$

$$F_{PO4} = 1 + c(Ca^{2+})\cdot f^{12}\cdot K_8 + c(Mg^{2+})\cdot f^{12}\cdot K_{15} \quad \text{(Eq 7.7-13)}$$

$$c(tH_2PO_4) = c(H_2PO_4^-)\cdot F_{H2PO4} \quad \text{(Eq 7.7-14)}$$

$$c(tHPO_4) = c(HPO_4^{2-})\cdot F_{HPO4} \quad \text{(Eq 7.7-15)}$$

$$c(tPO_4) = c(PO_4^{3-})\cdot F_{PO4} \quad \text{(Eq 7.7-16)}$$

7.8 Summary of the required analysis data

Analytical values

Full analysis - for ionic strength and accurate calculations

t_e Sampling temperature in °C

t_b Evaluation temperature in °C

t_{pH} Temperature at pH measurement in °C

pH

$K_{S4.3}$

$K_{B8.2}$ / t_{tit}

2 parameters required, for possible combinations see chap. 4.5.2

or alternatively:

TIC and pH

Ionic strength (calculated from full analysis)

alternatively:

Conductivity (25 °C) (for calculation of ionic strength)

In the open system additionally:

c_G CO_2 concentration in the gas phase

8 Evaluation variables for the carbonic acid system

In order to characterize the deviation of a water from the lime-carbonic acid equilibrium, certain evaluation variables have been created (Table 7.8-1).

Table 7.8-1: Evaluation variables for the carbonic acid system

pH value of calcite saturation (adjusted with calcite)	pH_C	
pH value of calcite saturation (adjusted with carbon dioxide)	pH_A	
pH value of calcite saturation (according to STROHECKER and LANGELIER)	pH_L	
Delta pH value	ΔpH	
Saturation index	SI	
Calcite dissolving capacity	D	$D > 0$
Calcite separation capacity	D	$D < 0$

The pH value of calcite saturation (saturation pH value) is established, when a water is in lime-carbonic acid equilibrium and thus neither lime dissolving nor lime depositing tendencies are present. In the state of calcite saturation applies:

$$pH = pH_A = pH_L = pH_C \quad \text{(Eq 7.8-1)}$$

as well as

$$SI = 0\ ;\ D = 0\ ;\ \Delta pH = 0 \quad \text{(Eq 7.8-2)}$$

During water treatment, the deacidification of water results in different pH values at which calcite saturation is reached, depending on the deacidification agent used (see also chapter 9.1.7). Therefore, different saturation pH values are distinguished depending on the process.

8.1 Saturation pH value after adjustment with calcite (pH_C)

The pH value of calcite saturation pH_C of a water is defined only for the case when the saturation with calcite has been set.

pH_C can be determined by practical tests (chap. 8.4.1) or calculated by a calculation program using iterative calculation steps. A statement on calcite saturation is obtained by comparing the pH value with the pH_C:

$pH < pH_C$: The water is calcite soluble

$pH = pH_C$: The water is in the state of calcite saturation

$pH > pH_C$: The water is calcite separating

In order to record the extent of the deviation of the pH value from the pH_C and to be able to indicate it as a value, it is useful to define the pH value difference as a delta pH value (ΔpH) (see chap. 8.4).

8.2 Saturation pH value after carbon dioxide exchange (pH_A)

The saturation pH value of a water, which occurs after the exchange of carbon dioxide (gas exchange process) in the water, is called pH_A.

The m-value remains unchanged with this procedure. (Chap. 7.2.3.1)

Also the calcium content of the water does not change by exchange of CO_2, as long as there is no precipitation of lime. Therefore, m-value and calcium content can be used as determinants of pH_A.

pH_A can be calculated using an iterative calculation program. Comparison of pH_A with the pH of the water gives an indication of calcite saturation:

$pH < pH_A$: The water is calcite dissolving

$pH = pH_A$: The water is in the state of calcite saturation

$pH > pH_A$: The water is calcite separating

8.3 Saturation pH value according to STROHECKER/LANGELIER (pH_L)

The saturation pH of a water, which would result without changing its calcium and hydrogen carbonate concentration **mathematically**, is called pH_L. This value was used rather historically.

If you form the equation for the 2. Equilibrium constant of carbonic acid, then with known activities of HCO_3^- and CO_3^{2-} the corresponding H^+ activity and thus the pH value of such a water can be calculated via the constant K_2:

$$\frac{a(CO_3^{2-}) \cdot a(H^+)}{a(HCO_3^-)} = K_2 \quad \rightarrow \quad a(H^+) = \frac{a(HCO_3^-)}{a(CO_3^{2-})} \cdot K_2 \qquad \text{(Eq 8.3-1)}$$

The activity of the carbonate ions in the calcite saturation state can be calculated by rearranging the definition equation for the solubility product.

$$K_C = a(Ca^{2+}) \cdot a(CO_3^{2-}) \qquad \text{(Eq 8.3-2)}$$

$$a(CO_3^{2-}) = \frac{K_C}{a(Ca^{2+})} \qquad \text{(Eq 8.3-3)}$$

Substituting this equation into (Eq. 8.3-1), the H^+ activity is calculated for the concentration of carbonate ions in the saturation state.

(Eq 8.3-3) in (Eq 8.3-1):

$$a(H^+) = \frac{K_2}{K_C} \cdot a(HCO_3^-) \cdot a(Ca^{2+}) \qquad \text{(Eq 8.3-4)}$$

The calculation is based on the actual amount of hydrogen carbonate ions of the initial water and the carbonate ion concentration calculated for saturation. One does not take into account that the amount of hydrogen carbonate ions would change due to the pH shift until saturation is reached and would then have a different value!

Transformation of (Eq. 8.3-4):

$$- \lg(a(H^+)) = - \lg(K_2) + \lg(K_C) - \lg(a(HCO_3^-)) - \lg(a(Ca^{2+})) \qquad \text{(Eq 8.3-5)}$$

Considering the definition of pH ($pH = - \lg(a(H^+))$), the equation according to STROHECKER and LANGELIER ($pH = pH_L$) is obtained.

$$pH_L = - \lg(K_2) + \lg(K_C) - \lg(a(HCO_3^-)) - \lg(a(Ca^{2+})) \qquad \text{(Eq 8.3-6)}$$

If one uses the concentrations of the compounds in calculations, then the activity coefficients must be taken into account. In sum, this equation gives an activity coefficient of f^5, where the negative decadic logarithm is used as (- 5·lg(f)).

$$pH_L = lg(K_C / K_2) - lg(c(HCO_3^-)) - lg(c(Ca^{2+})) - 5 \cdot lg(f) \qquad \text{(Eq 8.3-7)}$$

Comparison of pH_L with the pH of the water gives an indication of calcite saturation:

$pH < pH_L$:	The water is calcite dissolving
$pH = pH_L$:	The water is in the state of calcite saturation
$pH > pH_L$:	The water is calcite separating

When the pH of the water is less than 9, pH_L has approximately the value of pH_A [52].

8.4 Delta pH value (ΔpH)

Delta pH is defined as the pH difference between the pH value (pH) of a water and the pH after calcite saturation with calcite has been adjusted (pH_C).

$$\Delta pH = pH - pH_C \qquad \text{(Eq 8.4-1)}$$

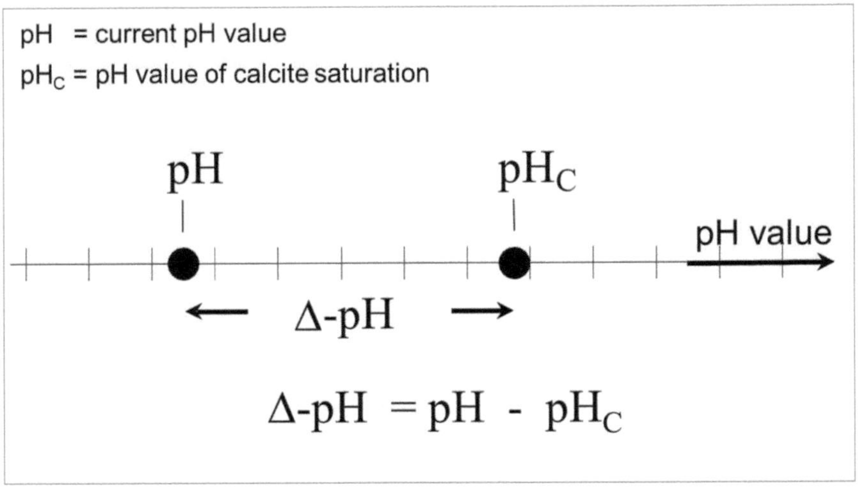

Figure 8.4-1: Definition of the ΔpH value

If the actual pH value is less than the saturation pH value

$$pH < pH_C$$

then lime can still dissolve in this water, the ΔpH value becomes negative !

Is $$pH > pH_C$$

so there is a lime supersaturated solution, the water is lime separating, the ΔpH-value becomes positive !

Is $$pH = pH_C$$

so neither lime is dissolved nor separated, the lime-carbonic acid equilibrium prevails.

Table 8.4-1: Overview of possible ΔpH values

Overview ΔpH values		
$pH < pH_C$	$\Delta pH = (-)$	lime dissolving
$pH > pH_C$	$\Delta pH = (+)$	lime separating
$pH = pH_C$	$\Delta pH = 0$	Lime-carbonic acid balance

In analogy to the calcite dissolving capacity (see below), a pH difference to pH_C was also defined as ΔDpH_C [51]:

$$\Delta DpH_C = pH_n - pH_v \qquad \text{(Eq 8.4-2)}$$

v = before calcite saturation
n = after calcite saturation

Compared to the above definition of ΔpH, there is a reversal of sign, but it is consistent with the calcite dissolving capacity (Table 8.4-2).

Table 8.4-2: Quantities describing the measure of calcite saturation

Quantities describing the measure of calcite saturation					
	ΔpH	SI	ΔDpH_C	D	
$pH < pH_C$	(-)	(-)	(+)	(+)	lime dissolving
$pH > pH_C$	(+)	(+)	(-)	(-)	lime separating
$pH = pH_C$	= 0	= 0	= 0	= 0	Calcite saturation

The sign analogy of ΔpH to the saturation index makes sense, which is why the use of ΔDpH_C should be avoided to avoid confusion.

According to the German Drinking Water Ordinance (TVO) of 1990, the pH of a drinking water was not allowed to be below the pH of calcite saturation, disregarding fluctuations of up to 0.2 pH units [73]. For the ΔpH value this meant:

Limit value according to TVO 1990: $\Delta pH \geq -0.2$

However, this limit is no longer current.

The ΔpH value can either be measured directly or, if the carbonic acid parameters are known, can be determined using a calculation program.

8.4.1 Measurement of the ΔpH value

Analytical detection of the ΔpH value is based on the principle that when water comes into contact with calcite, the pH, calcium content, total inorganic carbon (TIC) and acid capacity change.

If the water is lime-soluble, the values of these parameters increase due to the dissolution of calcite. If it is calcareous, the values will drop due to precipitation of calcite.

This process can be influenced by the presence of crystallization inhibitors such as humic substances or even polyphosphates. Comparison of measured values and calculated values for pH_C can give indications of such substances.

In DIN 38404-10 (1995 edition) [52] it was noted that this method was not suitable for determining the pH_C. A calculation method should be used to determine pH_C.

<u>**Measuring apparatus for on-site measurement**</u>

The crucial requirement for an apparatus for on-site measurement of the ΔpH value is that the water temperature must be kept constant over the measurement time and measured under exclusion of air. (Experimental setup analogous to [74])

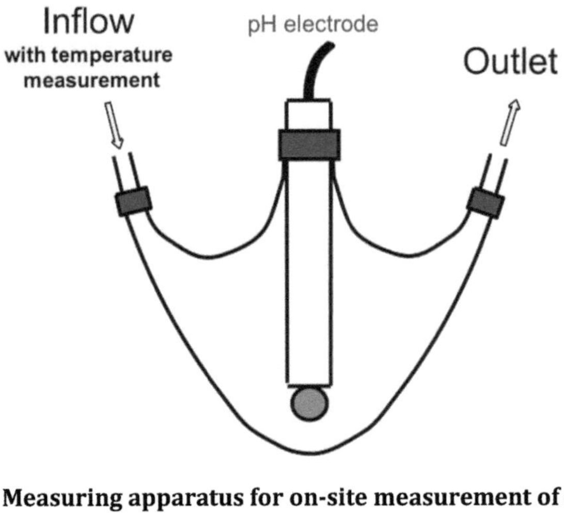

Figure 8.4-2: Measuring apparatus for on-site measurement of the ΔpH value

A pH electrode is inserted into a three-neck flask with inlet and outlet (Figure 8.4-2) so that the tip of the electrode is just above the bottom of the flask. In the inlet, the temperature of the water is measured with the temperature sensor of the pH device. The effluent is directed into the water bath in which the flask is immersed, so that the temperature of the water in the flask does not change during the test period.

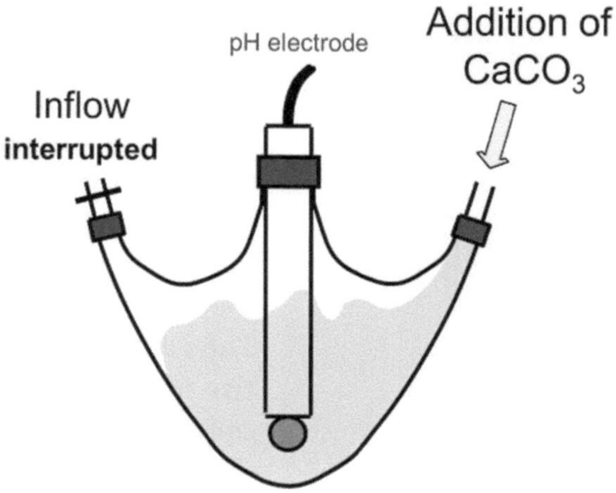

Figure 8.4-3: Carrying out the ΔpH measurement

The sample water is passed through the flask until the temperature and pH value remain constant. Then the water inlet is closed. (Experience has shown that the pH value can increase by up to 0.02 units if the flow is omitted)

The pH value is read.

Immediately afterwards, fine crystalline $CaCO_3$ powder is added through the outlet opening (Figure 8.4-3:)(a weighing boat made of glass with a long neck has proven to be useful here) until the pH electrode is surrounded by a cloud of $CaCO_3$ powder.

If the pH value then changes, the water was not in lime-carbonic acid equilibrium.

If it rises, the water was lime-dissolving; if it falls, it was lime-separating. After 2 minutes, read the pH value of the calcite saturation and calculate the ΔpH value.

Continuous measurement of the ΔpH value

To control water treatment processes, it is useful to continuously measure the ΔpH value. Flow meters are used for this purpose.

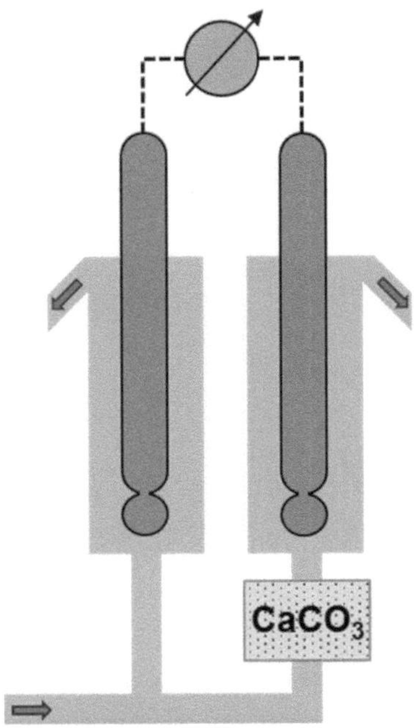

There are two pH electrodes in these devices. One is flowed around by untreated water and measures the initial pH value. The other electrode measures the pH in the water that has passed through a cartridge filled with $CaCO_3$ and is therefore saturated with lime. The difference between the two measured values gives the ΔpH value. This value is displayed directly.

Figure 8.4-4: Scheme of continuous measurement of the ΔpH value

Determination of calcite saturation by marble dissolution test

The DIN 38404-10 (1995 edition) [52] provided for a marble dissolution test to determine the calcite saturation.

To determine the calcite saturation by marble dissolution test, the change in pH and acid capacity 4.3 are determined by the saturation of water with calcite.

Two bottles (250 ml) are filled with the sample water. To exclude contact of the sample water with the ambient air, make sure that the filling hose reaches the bottom of the bottle and that the sample water overflows for about one minute and the bottle is closed without air bubbles.

To keep the temperature constant, both bottles are placed in a water bath. If the determination is to be carried out on site, the water bath is flown through by water from the sampling tap. If the measurement is to be carried out in the laboratory, then the bottles are transported in a cool bag and the water bath is cooled via a thermostat and the temperature is kept constant to ±1 K.

A magnetic stir bar and 2 to 3 g of calcite powder are added to the first bottle. The stirring time is two hours. After a further waiting time of 5 minutes, pH, temperature and $K_{S4.3}$ are determined in both bottles.

The pH value measured in the bottle with calcite powder is the pH_C value, determined by calcite dissolution test.

In addition, the pH_C value should be determined by calculation. If the calculated pH_C value is higher than that determined by marble dissolution testing, this is an indication of the presence of crystallization inhibitors.

The calcite dissolving capacity of the water can be determined via the calcite dissolution test (see chapter 8.6).

8.5 Saturation index SI

The saturation index SI is purely a calculated value and cannot be determined by direct analysis. It is used to assess the behavior of water against lime.

In contrast, the ΔpH value, which is also used to assess the behavior of water with respect to lime, is a measured value that indicates the deviation of the actual pH value from the pH value of calcite saturation (lime-carbonic acid equilibrium). The amount of the saturation index is always greater than that of the ΔpH value. The signs correspond (solution of lime: (-) ; deposition of lime: (+))!

The saturation index gives the deviation of the product of the actual activities of Ca^{2+} and CO_3^{2-} from the theoretical value of the solubility product $K_C(CaCO_3)$.

Definition:

$$SI = \lg \frac{a(Ca^{2+}) \cdot a(CO_3^{2-})}{K_C} \qquad \text{(Eq 8.5-1)}$$

$$K_C = a(Ca^{2+}) \cdot a(CO_3^{2-}) \quad \text{(at the solubility limit!)} \qquad \text{(Eq 8.5-2)}$$

Is the product of the activities

$a(Ca^{2+}) \cdot a(CO_3^{2-}) > K_C$ Solubility of calcite exceeded, SI is positive **(+)**
➔ the water is lime separating !

$a(Ca^{2+}) \cdot a(CO_3^{2-}) < K_C$ $CaCO_3$ can still dissolve, SI is negative **(-)**
➔ the water is lime dissolving !

$a(Ca^{2+}) \cdot a(CO_3^{2-}) = K_C$ then SI = **0**
➔ the water is in balance!

The concentration of the carbonate cannot be determined analytically with sufficient accuracy and must be calculated via the concentration of the hydrogen carbonate. The definition equation must therefore be rearranged accordingly for the calculation.

The CO_3^{2-} activity is calculated using the second mass action equation.

$$\frac{a(CO_3^{2-}) \cdot a(H^+)}{a(HCO_3^-)} = K_2 \qquad \text{(Eq 8.5-3)}$$

$$\rightarrow \quad a(CO_3^{2-}) = \frac{a(HCO_3^-) \cdot K_2}{a(H^+)} \qquad \text{(Eq 8.5-4)}$$

(Eq 8.5-4) in (Eq 8.5-1): $$SI = \lg\left(\frac{a(Ca^{2+})\cdot a(HCO_3^-)\cdot K_2}{K_C\cdot a(H^+)}\right)$$ (Eq 8.5-5)

Since the analysis determines the substance amount concentrations, the ionic strength must be taken into account:

$$SI = \lg\left(\frac{c(Ca^{2+})\cdot f^4\cdot c(HCO_3^-)\cdot f\cdot K_2}{K_C\cdot a(H^+)}\right)$$ (Eq 8.5-6)

It follows

$$SI = \lg\left(\frac{c(Ca^{2+})\cdot c(HCO_3^-)\cdot K_2}{K_C\cdot a(H^+)}\cdot f^5\right)$$ (Eq 8.5-7)

Annotation: Since the activity of the H^+ ions is recorded directly with the pH electrode, $a(H^+)$ can be derived directly from the pH value and does not have to be taken into account when calculating the activity coefficient!

The saturation index can also be calculated with the pH value according to STROHECKER and LANGELIER pH_L (see chap. 8.3):

$$SI = pH - pH_L$$ (Eq 8.5-8)

Derivation of the formulaic relationships:

Transform equation 8.5-6:

$$SI = -\lg(a(H^+)) + \lg\{c(Ca^{2+})\cdot c(HCO_3^-)\cdot(K_2/K_C)\cdot f^5\}$$ (Eq 8.5-9)

with $pH=-\lg(a(H^+))$: ➔ $SI=pH+\lg\{c(Ca^{2+})\cdot c(HCO_3^-)\cdot(K_2/K_C)\cdot f^5\}$ (Eq 8.5-10)

Definition equation for pH_L:

(Eq 8.3-7) $pH_L=-\lg(K_2)+\lg(K_C)-\lg(c(Ca^{2+}))-\lg(c(HCO_3^-))-5\cdot\lg(f)$ (Eq 8.5-11)

Transform: $-pH_L = \lg\{c(Ca^{2+})\cdot c(HCO_3^-)\cdot(K_2/K_C)\cdot f^5\}$ (Eq 8.5-12)

(Eq 8.5-12) in (Eq 8.5-10): ➔ $SI = pH - pH_L$ (Eq 8.5-13)

8.6 Calcite dissolving capacity and calcite separation capacity

Calcite dissolving capacity D is the amount of substance or mass of calcite that a calcite dissolving water is capable of dissolving per liter. The calcite dissolution capacity can be calculated with a calculation program or determined practically with the calcite dissolution test (see chapter 8.4.1).

In its true sense, calcite dissolving capacity is defined by the difference in the calcium content of a calcite dissolving water before and after the addition of calcite.

$$D = c(Ca)_n - c(Ca)_v \qquad \text{(Eq 8.6-1)}$$

$( )_v$ = before addition of calcite
$( )_n$ = after reaction with calcite

However, one can equally define:

$$D = \tfrac{1}{2}\,(K_{S4,3})_n - \tfrac{1}{2}\,(K_{S4,3})_v \qquad \text{(Eq 8.6-2)}$$

If one equates the acid capacity with the m-value and considers the definition of the m-value (chap. 7.2.3.1) via the charge balance, it becomes clear that half of the change in the acid capacity (the m-value) corresponds to the change in the Ca ion concentration.

For the definition of the calcite dissolving capacity via the acid capacities, the same prerequisite applies as for the determination of acid and base capacities: Weak acids and bases other than the carbonic acid forms may be present in low concentrations at most !

If in Equation 8.6-1 $c(Ca)n < c(Ca)v$, then the concentration of calcite in the water has decreased by precipitation (lime deposition). The result of the difference becomes negative ($D < 0$). This is then referred to as the calcite separation capacity.

Calcite separation capacity D is the amount of substance or mass of calcite that a calcite-precipitating water can separate per liter.

Table 8.6-1: Calcite dissolving capacity and calcite separating capacity D

Calcite dissolving capacity	$D > 0$
Calcite separation capacity	$D < 0$

In the German Drinking Water Ordinance of May 2001 (TVO) [75] the calcite dissolving capacity was introduced as a limit value and replaced the previously applicable limit value for the ΔpH value.

Limit value according to german TVO : $D \leq 5$ mg/L

At a pH ≥ 7.7, this limit value no longer has to be taken into account in accordance with the Drinking Water Ordinance.

9 Water treatment

9.1 Deacidification in drinking water treatment

9.1.1 Legal regulations

Originally, deacidification was generally carried out for reasons of corrosion protection without any legal requirement to do so. Since 1986, however, deacidification has been mandatory in certain cases under the German Drinking Water Ordinance. The background to the introduction of a limit value in the Drinking Water Ordinance was the fear that, in the case of fiber cement materials, fibers could dissolve in water with a high lime dissolving content and enter the drinking water.

This originally purely technical problem also acquired health relevance through reports that asbestos fibers, which can be washed out of fiber cement pipes by lime-dissolving waters, can also cause cancer in the gastrointestinal tract if ingested orally. This statement is still debated among experts [76] [77].

For precautionary reasons, however, a limit value was newly included in the German Drinking Water Ordinance in 1986 [78]. According to this, the pH value of the discharged water must not be below the pH value of the calcium carbonate saturation for metallic or cementitious materials in the pH range 6.5 - 8.0 , and for fiber cement materials in the pH range 6.5 - 9.5. In a note, it was remarked that fluctuations of the pH of the water below the pH of calcium carbonate saturation up to 0.2 pH units were disregarded.

To represent compliance with the limit, the difference between pH_C and pH was used as the ΔpH value (see 8.4), which subsequently had to be greater than or equal to -0.2.

$$\Delta pH = pH - pH_C \geq -0.2 \qquad \text{(Eq 9.1-1)}$$

This limit was changed in subsequent revisions of the German Drinking Water Ordinance.

The currently valid limit value (2022) of the German Drinking Water Ordinance is that the calculated calcite dissolving capacity at the outlet of the waterworks must not exceed 5 mg/L $CaCO_3$. This requirement is considered to be met when the pH at the waterworks outlet ≥ is 7.7 [75].

Table 9.1-1: Relevant limit values according to Annex 3 of the TVO 2001 [75]

Hydrogen ion concentration	pH units	≥ 6.5 and ≤ 9.5
Calcite dissolving capacity D	mg/L $CaCO_3$	5

Thus, if the pH value at the waterworks outlet is ≥ 7.7, the limit value is considered to be complied with. This means that a determination of the calcite dissolving capacity is then not necessary.

When mixing drinking waters in the distribution network, the calcite dissolving capacity must not exceed 10 mg/L.

If the limit value cannot be met, the water must be deacidified. When assessing whether a drinking water needs to be deacidified, a decision scheme can help (Figure 9.1-1:).

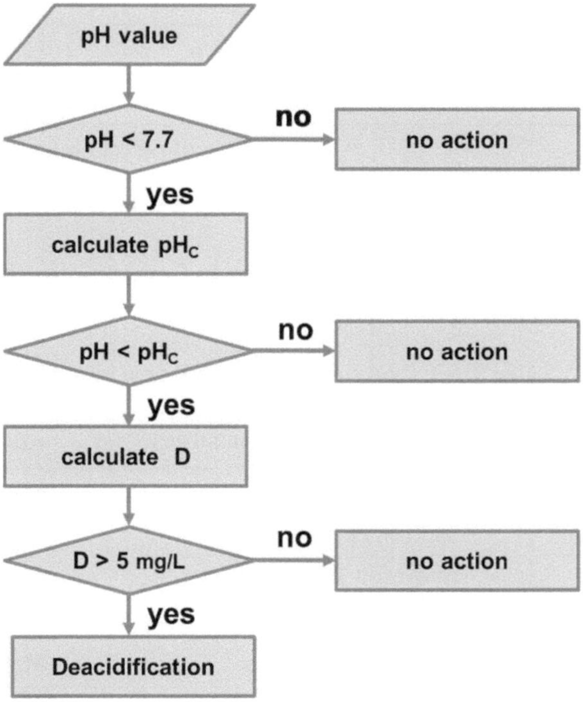

Figure 9.1-1: Decision scheme deacidification

9.1.2 Basics of deacidification

Waters that need to be deacidified are **lime-dissolving** and have a **pH value that is too low** in relation to the lime-carbonic acid balance.

$$pH < pH_C$$

According to their purpose, deacidification processes therefore always lead to an increase in the pH value of the water.

Definition of deacidification:

Increase of the pH value to achieve calcite saturation

In terms of the process engineering principle, there are two basic ways to deacidify a water (Table 9.1-2:).

Table 9.1-2: Overview of the deacidification process technology

Process engineering principle	Procedure
Physical deacidification	- Gas exchange process
Chemical deacidification	- Dosing method (chemical dosing) - Filter process (deacidification filter)

The agents listed in Table 9.1-3: are approved in Germany for the deacidification of drinking water [79].

Table 9.1-3: Overview of approved deacidifiers

Gas exchange	Carbon dioxide (discharge)	- CO_2
Dosing method	Caustic soda	$NaOH$
	Sodium carbonate = soda	Na_2CO_3
	Hydrated lime	$Ca(OH)_2$
Filter method	Limestone (marble)	$CaCO_3$
	semi-fired dolomite	MgO / $CaCO_3$

9.1.3 Physical deacidification

The goal of physical deacidification is to increase pH by removing carbonic acid ($-CO_2$) through gas exchange with air.

Since the gas exchange is carried out open to the atmosphere, the water is always depressurized after treatment and must then be pumped further.

Process engineering principle of gas exchange

All gas exchange processes strive towards a saturation concentration, which is described by HENRY-DALTON's law (see chapter 4.1).

The BUNSEN absorption coefficient α (see chapter 4.1) directly describes the relationship between the CO_2 concentration in water and the concentration in air.

$$\frac{c_W}{c_G} = \text{constant} = \alpha \qquad \text{(Eq 9.1-2)}$$

c_W : Molar concentration in water
c_G : Molar concentration in the gas

The BUNSEN absorption coefficient for CO_2 for 10 °C ($\alpha(CO_2) = 1.205$) shows that CO_2 is very soluble in water. For comparison, oxygen is poorly soluble with a coefficient of $\alpha(O_2) = 0.038$.

However, due to the good solubility of CO_2 in water, it is also relatively difficult to remove it from the water again. This can be illustrated with the help of a bottle experiment for the gas discharge of CO_2 in exchange with the atmosphere.

Bottle test

One bottle is half filled with water, the other half is ambient air. The CO_2 concentration in the water is assumed to be 40 mg/L. Shake the closed bottle until equilibrium is reached (Figure 9.1-2:).

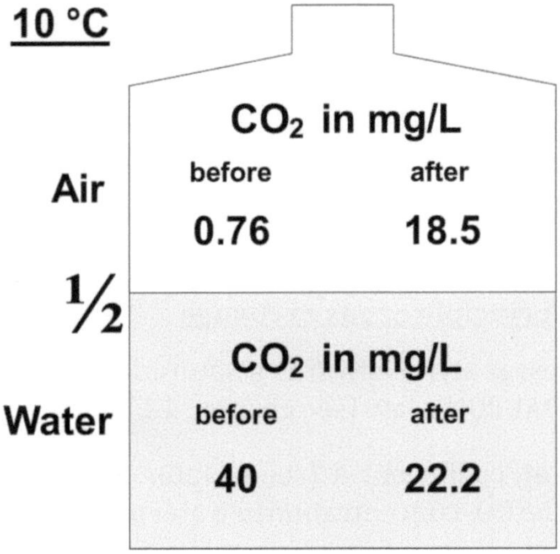

Figure 9.1-2: Bottle experiment for CO_2 discharge from water

After shaking, the CO_2 concentration in the water has almost halved. CO_2, however, has accumulated strongly in the air.

If one wanted to reduce the CO_2 concentration in the water more strongly, one would have to replace the air in the bottle with fresh air and repeat the process (Figure 9.1-3:).

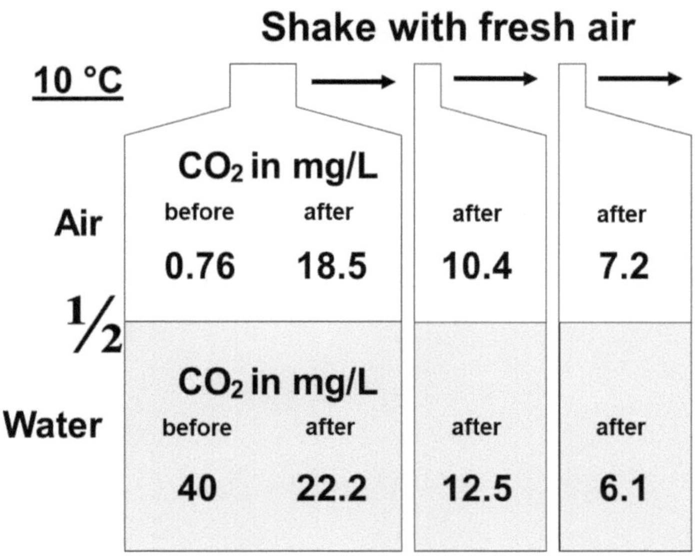

Figure 9.1-3: Bottle experiment for multi-stage CO_2 discharge from water

But again, you would only about halve the CO_2 concentration of the water to 12.5 mg/L. If one wanted to achieve complete equilibrium with the air (equilibrium concentration 0.99 mg/L), one would have to repeat the process several times.

Considering raw waters for drinking water treatment, groundwaters are enriched with CO_2 due to soil passage during rainwater infiltration (Section 5.4). In the case of reservoir waters and lakes, the water is usually taken from the deep layer (hypolimnion), where CO_2 has been enriched in the water due to the predominance of degradation processes (chap. 5.7). Both types of raw water for drinking water treatment are therefore enriched with CO_2 compared to the atmosphere (ambient air). In contact with air, CO_2 will therefore escape from the water, since the gas solubility in the depressurized water is exceeded. It is therefore possible to extract CO_2 from such water by gas exchange with the ambient air and thus to deacidify it.

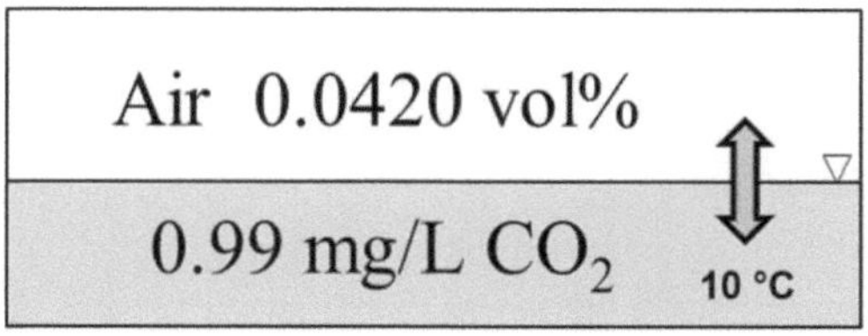

Figure 9.1-4: Scheme of distribution of CO_2 between air and water

However, CO_2 can be removed from water in gas exchange with ambient air to a maximum of the equilibrium concentration of gas solubility (0.99 mg/L at 10 °C, see 4.1)(Figure 9.1-4). Then the system is in equilibrium and no more exchange against the ambient air is possible.

This theoretical upper limit is reached at different pH values depending on the initial concentration of CO_2 or TIC, as can be seen from Table 9.1-4:.

Table 9.1-4: Upper limit of physical deacidification (calculated values, 10 °C)

	TIC in mmol/L					
	0.5	1	2	3	4	5
pH	CO_2 concentration in water mg/L					
9.0	0.062	0.12	0.25	0.37	0.50	0.62
8.9	0.079	0.16	0.31	0.47	0.63	0.79
8.8	0.099	0.20	0.40	0.60	0.79	0.99
8.7	0.13	0.25	0.50	0.75	1.0	1.3
8.6	0.16	0.32	0.63	0.95	1.3	1.6
8.5	0.20	0.40	0.80	1.2	1.6	2.0
8.4	0.25	0.50	1.0	1.5	2.0	2.5
8.3	0.32	0.63	1.3	1.9	2.5	3.2
8.2	0.40	0.79	1.6	2.4	3.2	4.0
8.1	0.50	0.99	2.0	3.0	4.0	5.0
8.0	0.62	1.2	2.5	3.7	5.0	6.2
7.9	0.78	1.6	3.1	4.7	6.2	7.8
7.8	0.97	1.9	3.9	5.8	7.8	9.7
7.7	1.2	2.4	4.8	7.3	9.7	12
7.6	1.5	3.0	6.0	9.0	12	15
7.5	1.9	3.7	7.4	11	15	19
7.4	2.3	4.6	9.2	14	18	23
7.3	2.8	5.6	11	17	22	28
7.2	3.4	6.8	14	21	27	34
7.1	4.1	8.3	17	25	33	41
7.0	5.0	9.9	20	30	40	50

Theoretical upper limit
Equilibrium concentration
with air 0.99 mg/L

Practical upper limit
Residual concentration 3 mg/L

·········· = upper limit of the theoretically possible gas exchange
— — — = residual concentration 3 mg/L CO_2
(box with diagonal) = pH value of calcite saturation is reached
(shaded) = possible range for gas exchange processes

In a very soft water with lower TIC values (TIC 0.5 mmol/L), this limit of gas exchange is already reached at pH 7.8. However, only at pH 8.9 would this water be in the range of calcite saturation. This water could not be deacidified by gas exchange.

Beyond this theoretical limit, there is also a practical limit up to which gas exchange can be carried out with a "reasonable" effort. It is reported to be about 3 mg/L residual CO_2 concentration [80].

With the very soft water described, the practically achievable pH value would then even be only 7.2. The pH_C could not be reached. Even in a water with 1 mmol/L TIC, calcite saturation could not be reached by gas exchange (pH 8.3), since the practical limit of 3 mg/L CO_2 is already reached at pH 7.6.

Generally, therefore, the **upper limit of** physical deacidification is given as **pH 7.8** . If the pH of calcite saturation is higher, it cannot be achieved by gas exchange alone.

This means that waters with TIC contents below 2 mmol/L (very soft waters) cannot be deacidified to calcite saturation using a gas exchange process. The reduction in buffer capacity associated with carbonic acid removal speaks against partial deacidification of these waters by gas exchange.

In harder waters (higher TIC values), the pH of calcite saturation is already reached below the practical upper limit (pH 7.7 at TIC of 2 mmol/L), so that gas exchange can be used here for deacidification.

Changes in water due to physical deacidification

Physical deacidification causes the following changes to occur in the water:

- p-value increases / (- p-value decreases !)
- m value remains unchanged (Chap. 7.2.3.1)
- the hardness of the water remains unchanged (Prerequisite: no lime precipitation)

Process engineering principle of physical deacidification

Since all gases must pass through the interface between water and air, one must always strive to make this interface as large as possible.

Interface magnification:

- fine water droplets in air
- fine air bubbles in water
- thin water films on surfaces
- Trickling on walls or packed beds

Requirements for gas exchange in practice:

- water/air interface as large as possible
- short dwell times
- high concentration gradient between water and air

It is true for all physical processes that an increasing efficiency causes an increasing energy demand.

In terms of process technology, a basic distinction is made between co-current processes, cross-current processes and counter-current processes, with the effectiveness increasing in the sequence.

The ideal **co-current process** corresponds to the bottle model. You bring the water into contact with a certain volume of air once and make sure that the interface becomes as large as possible (shake). In terms of process technology, there are no ideal pure DC processes.

In the **cross-flow process** the process of gas exchange by cross-flow is repeated several times in the same body of water and the air is renewed each time. The concentration in the water decreases in the direction of water flow. The air requirement increases sharply compared to the DC process.

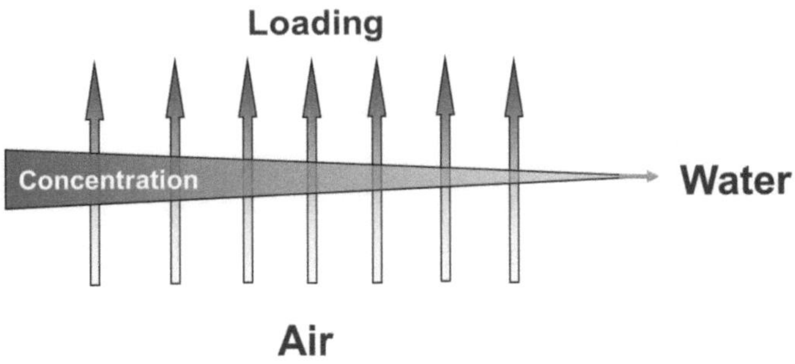

Figure 9.1-5: Scheme cross flow method

In **counter-current systems** the air is guided against the water flow. The cleanest air comes into contact with low-polluted water. The air, which is already somewhat contaminated afterwards, comes into contact with more highly contaminated water, so that there is always a sufficient concentration gradient.

Countercurrent systems reduce the air requirement (energy demand) compared to crosscurrent systems, but are more complex to build.

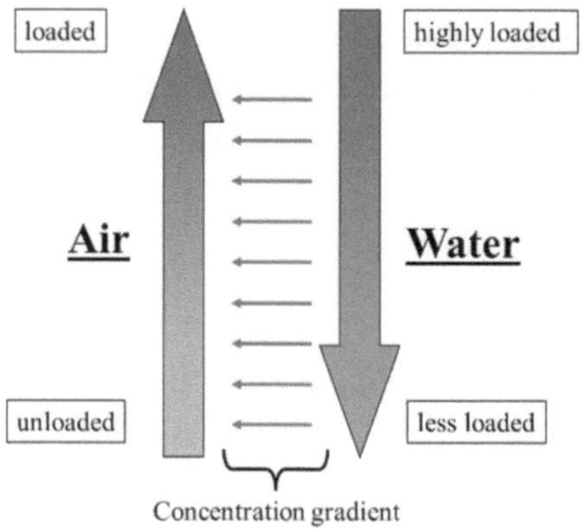

Figure 9.1-6: Scheme counterflow method

Process engineering

Atomization

A frequently used process is the atomization of the water.

Target: droplets as small as possible = large surface area

In a reaction chamber, the water is sprayed under a pressure of 0.5 to 2 bar to form fine droplets of less than 1 mm in diameter, bringing them into contact with the air over a large area.

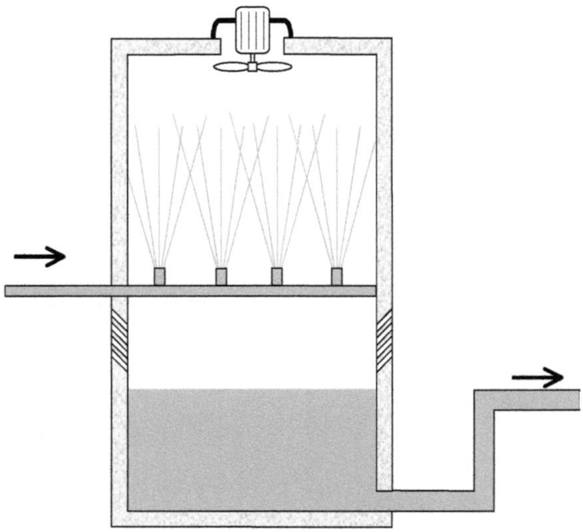

Figure 9.1-7: Atomization plant (schematic)

A suitable arrangement of the nozzles is aimed at ensuring that the sprayed water remains in the air for as long as possible. The water-saturated atmosphere in the reaction chamber is very corrosive, so only resistant materials can be used.

Cascades

Technically particularly simple is the gas exchange via cascades, where in the simplest case the water just falls down a staircase and experiences an air exchange from step to step.

With a sufficient number of stages (e.g. about 10), at least as good an effect is achieved with regard to CO_2 outgassing as with atomization.

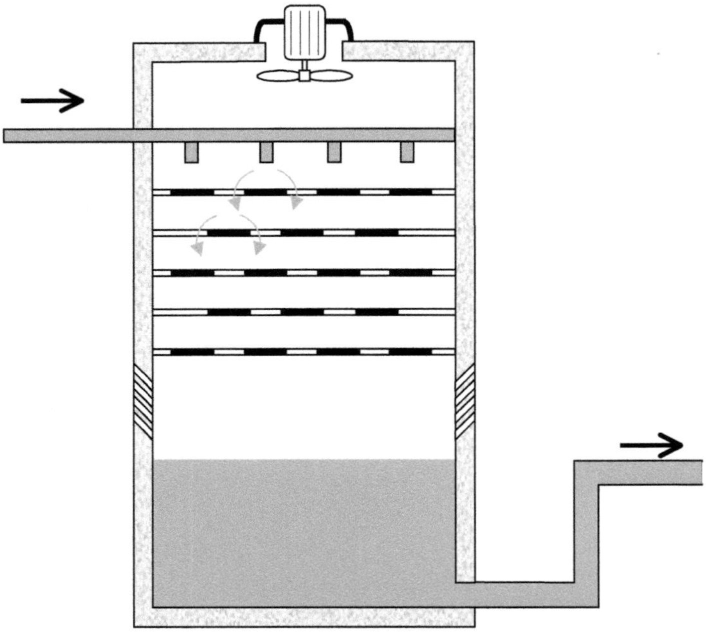

Figure 9.1-8: Cascade ventilation (schematic)

Figure 9.1-9: Photo of a cascade

Bubble ventilation

Bubble aeration involves the introduction of small air bubbles into the water. In doing so, one can apply an arbitrarily high excess of air and thus approach the air saturation equilibrium (0.99 mg/L) also with respect to CO_2 outgassing at high energy input.

Trickle film ventilation

The goal is to achieve a large surface area by producing thin liquid films.

The technical design is carried out as so-called corrugated sheet columns made of very thin corrugated plastic sheets, which are suspended at intervals of a few cm in appropriate shafts.

Figure 9.1-10: Corrugated sheet columns

<u>Pressurized ventilation - avoidance of CO_2 discharge</u>

During gas exchange for **oxygen enrichment**, CO_2 is also always discharged if it is above the saturation concentration with the air.

This may be undesirable if the water is already in lime-carbonic acid equilibrium and lime precipitation may then occur.

In this case, pressure ventilation can then be selected, where the partial pressure of the CO_2 is so high that outgassing cannot occur.

Advantage gas exchange process:	No hardening
Disadvantage gas exchange process:	Only applicable up to pH 7.8

9.1.4 Chemical deacidification

A distinction is made between:
- Dosing method (chemical dosing)
- Filter process (use of deacidification filters)

9.1.4.1 Dosing method

Characteristic of all dosing processes:

- good adaptation to raw water flow rates
- relatively low specific area and space requirements

However, these processes can only be used sensibly with appropriate measurement and control technology.

It must be measured:

1. Raw water delivery rate
2. pH value before deacidification
3. pH value after deacidification
4. ΔpH value

These systems are often controlled by the ΔpH value.

Chemicals for dosing:	Caustic soda	$NaOH$
	Sodium carbonate = soda	Na_2CO_3
	Hydrated lime	$Ca(OH)_2$

Deacidification with sodium hydroxide solution

When dosing sodium hydroxide solution (usually 50%), immediate good mixing is required, otherwise strong local pH increase and lime precipitation may occur ⟶ dilute again if necessary!

Technical facilities:

- Reservoir
- Dosing pump
- Dosing point (possibly with static mixer)

Advantage: No hardening !

Disadvantage: Increasing sodium concentration is disliked for health reasons. Limit value German TVO 2001: Na 200 mg/L

Deacidification with soda (Na_2CO_3)

Technical facilities:

- Storage room (bagged goods)
- Dissolving tank
- Dosing tank
- Dosing pump
- Dosing point (possibly with static mixer)

Advantage: No hardening !
Increase of buffering intensity (CO_3^{2-})

Disadvantage: Increasing sodium concentration is disliked for health reasons. Limit value German TVO 2001: Na 200 mg/L

Deacidification with hydrated lime ($Ca(OH)_2$)

Unslaked lime (CaO) or slaked lime ($Ca(OH)_2$) (hydrated lime) can be used for deacidification (for forms of lime see 4.7). Unslaked lime must then be slaked directly on site. The advantage of unslaked lime is the generally lower procurement costs, but this is offset by additional equipment costs and increased safety requirements.

Since the lime used is usually a natural product and therefore contaminated, cleaning or clarification is necessary when using it for drinking water treatment. From the lime hydrate, a lime milk is first prepared, which, depending on the requirements, can be dosed directly or transferred to a clarified and saturated lime water via a lime saturator (Figure 9.1-11:).

Milk of lime is a suspension (turbid), lime water is a real solution (clear).

When hydrated lime is dosed upstream of a filter stage in drinking water treatment, it can be dosed as milk of lime because the impurities are filtered off. When dosing into drinking water, clear lime water must be used.

Technical facilities:

- Storage silo
- Metering screws
- Lime milk settling tank
- Lime saturator ($\Rightarrow$ lime water)
- Lime water pumps
- Dosing point

In Figure 9.1-11: a lime saturator for the production of lime water is schematically shown. The widening of the cross-section results in the slowing of the rate of ascent, allowing contaminants to sediment. The sludge formed has a very high pH, this must be taken into account during disposal/use.

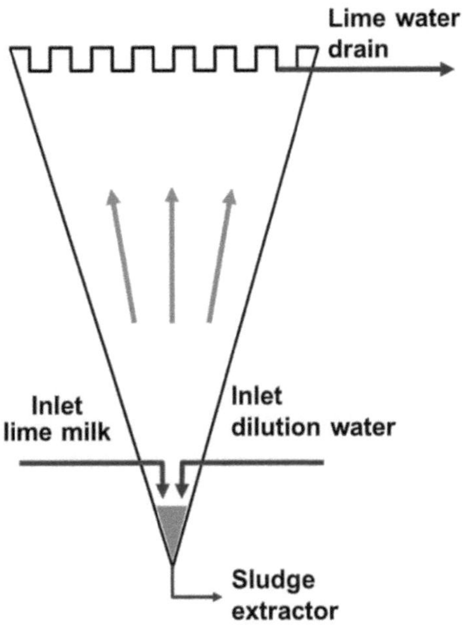

Figure 9.1-11: Schematic representation of a lime saturator

The lime water is free of carbonates, as these are completely precipitated and sedimented during the clarification process of the rising lime water due to the high pH value. When the lime water at the surface comes into contact with the CO_2 in the air, calcite precipitates again, forming a layer on the surface (Figure 9.1-12:).

Figure 9.1-12: Surface of a lime saturator with calcite precipitation

Advantage:	Hardening makes sense for soft water
Disadvantage:	Increase the hardness of hard water

9.1.4.2 Filter process

A characteristic feature of the deacidification filter processes is that structural equipment is used which corresponds in design to a filter for turbidity removal.

The filter material used is then a chemically active mass that reacts with the water or water constituents. The filter material is constantly consumed and must be refilled each time. After consumption of approx. 10% of the nominal quantity, refilling should be carried out each time.

Deacidification filters naturally also have a filtering effect. Care should therefore be taken to ensure that turbidity, iron and manganese concentrations are not too high. Otherwise there is a risk that the filter material will be inactivated by excessive deposits.

Common filter materials for deacidification:

Limestone	$CaCO_3$
semi-fired dolomite	$CaCO_3$ / MgO

Limestone $CaCO_3$

Limestone is a natural product and is sold under various trade names.

Deacidification with limestone is a **pure dissolving process**.

Decrease in reaction speed

- with increasing carbon dioxide content
- with increasing hardness

The use of this filter material should therefore be limited to waters with low hardness levels. The increase in hardness associated with this deacidification is also useful for soft waters. The further hardening and the decrease of the reaction rate argue against the application of the process for harder waters.

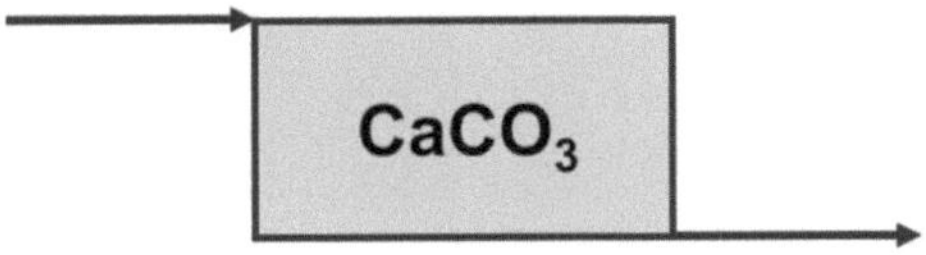

Figure 9.1-13: Scheme deacidification filtration with calcite

The control takes place via the filter speed (contact time). If the filtration rate is too low, there are no disadvantages - the solubility product cannot be exceeded and no lime can precipitate. If the residence time in the filter is too short and the lime-carbonic acid equilibrium is not reached as a result, post-deacidification may be necessary. It then makes sense to use a simple dosing method, e.g. NaOH.

Advantage: Hardening makes sense for soft water
Lime can not precipitate

Disadvantage: Increase the hardness of hard water
Decrease in reaction rate with increasing hardness
Only suitable for soft waters

Semi-fired dolomite

Dolomite	$CaCO_3 / MgCO_3$
Semi-fired	$CaCO_3 / MgCO_3 \rightarrow CaCO_3 / MgO + CO_2 \nearrow$
Reaction with water	$MgO + H_2O \rightarrow Mg^{2+} + 2\ OH^-$

Semi-fired dolomite is more reactive than limestone, since at the same time as the lime solution is formed, the pH is also raised by the hydroxide formed.

Figure 9.1-14: Scheme deacidification filtration with semi-fired dolomite

It depends on the filter speed and thus on the contact time of the water, which pH value is reached. If the contact time is too long (filter speed too low), the pH value of the lime-carbonic acid equilibrium is exceeded and precipitation of lime occurs ⟶ risk of caking of the filter material.

When the filter is not in use, a very high pH value is established in the filter bed. Commissioning must then be preceded by flushing, or the first filtrate must be discarded. Even after the filling of new material, easily soluble components can lead to higher pH values.

The pH value of the water is controlled by the filter speed v_f. In case of flow rate fluctuations, filters must be taken out of operation or additionally put into operation.

Advantage: Hardening makes sense for soft water

Disadvantage: Increase the hardness of hard water
Filter velocity v_f must be maintained very precisely
Lime can precipitate

9.1.5 Examples of process variants

General criteria for the selection of the deacidification process

- Hard waters should not be further hardened
- very soft waters should be hardened if possible → Increase of buffer capacity

➢ Very soft water with approx. 0,4 mmol/L hardness (2 °dH)

→ Filtration over limestone
- possibly post treatment deacidification with NaOH
- possibly **additional** hardening with CO_2 dosing before the filter stage

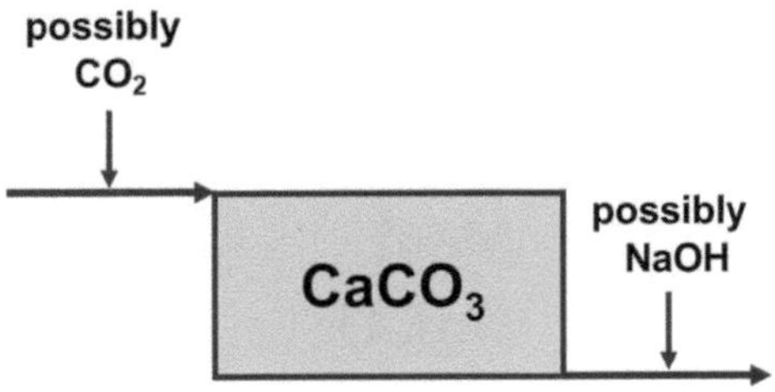

Figure 9.1-15: Scheme: Deacidification filter with pre- and post-treatment

- Very soft water - two-stage treatment with autocatalytic demanganization
  - ➔ Deacidification with hydrated lime

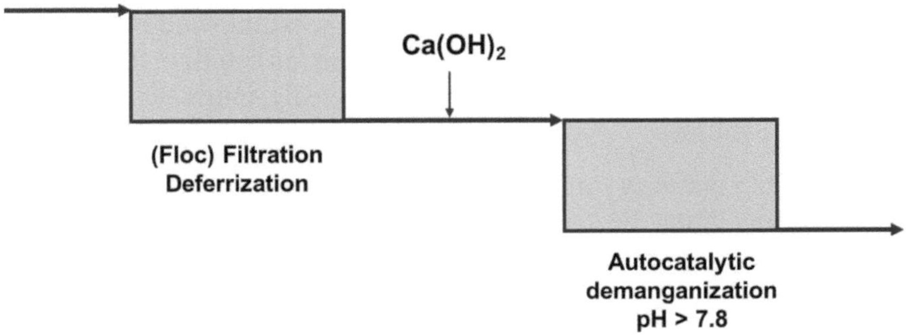

Figure 9.1-16: Scheme: Deacidification with hydrated lime

➔ no deacidification via gas exchange possible ! pH > 7.8 !
(typical treatment for soft reservoir waters)

- Water with total hardness approx. 1,8 mmol/L (8 °dH)
  - ➔ Deacidification through dosage of NaOH

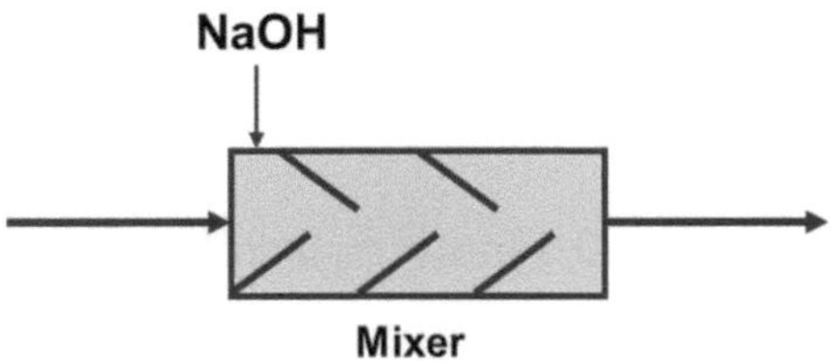

Figure 9.1-17: Scheme: Deacidification with caustic soda

- Hard water with e.g. 4,3 mmol/L hardness (24 °dH)
    - ➔ Deacidification by CO_2 removal through gas exchange

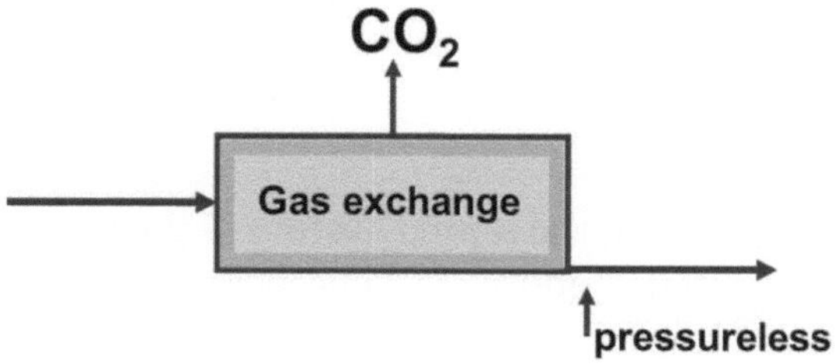

Figure 9.1-18: Scheme: Deacidification through gas exchange

9.1.6 Chemical reactions in the deacidification processes

Based on the reaction scheme described in chapter 4.7.2, the chemical reactions in the different deacidification processes can be made clear. In addition, it is possible to understand that the pH of the lime-carbonic acid equilibrium achieved after treatment varies for the same water depending on the deacidification process.

Since there is usually no lime in the piping system of the water treatment plants, the $CaCO_3$ was only indicated shadowy in the reaction scheme (Figure 9.1-19).

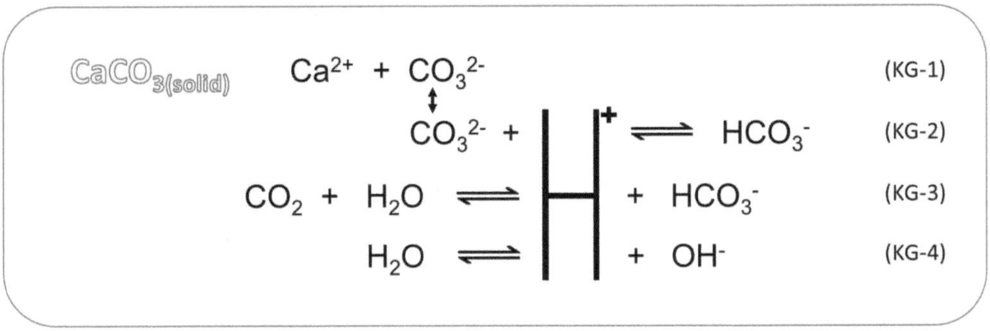

Figure 9.1-19: Carbonic acid scheme (without calcite)

➢ **Physical deacidification: CO_2 removal**

Physical deacidification is caused by the removal of CO_2. The loss of CO_2 shifts KG-3 to the left, H^+ is consumed, and the pH increases. The KG-2 and KG-4 reactions partially compensate for the H^+ loss. KG-2 increases the carbonate concentration.

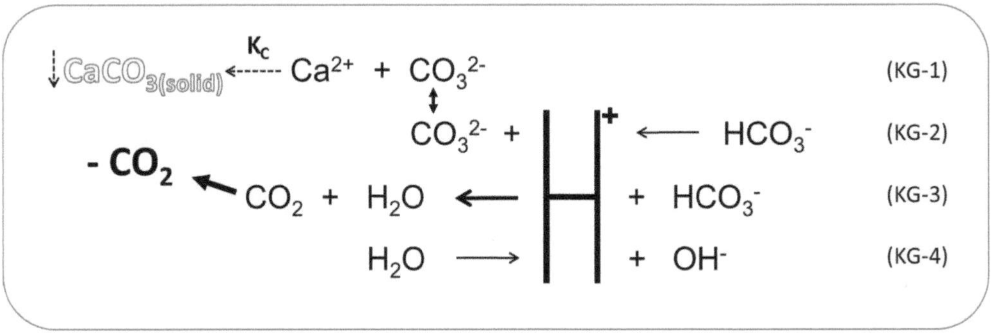

Figure 9.1-20: Physical deacidification with CO_2 removal

When enough carbonate is formed to reach the solubility product for $CaCO_3$, the goal of deacidification is achieved. The activity $a(H^+)$ then present in the water corresponds to the pH_A.

If CO_2 continues to be removed, calcite saturation will be exceeded, the water will become calcareous and lime may precipitate.

- Lime can precipitate if CO_2 removal is too strong
- The water hardness remains unchanged

➢ <u>Chemical deacidification / dosing process / caustic soda:</u>

$$NaOH \rightleftharpoons Na + OH^-$$

The components effective for deacidification are the hydroxide ions. The added hydroxide ions shift KG-4 to the left, consuming H^+ and increasing the pH. KG-2 and KG-3 partially compensate for the H^+ loss.

KG-2 increases the carbonate concentration until calcite saturation is reached. If caustic soda continues to be dosed, the solubility product is exceeded by the further increase in carbonate concentration and lime can precipitate.

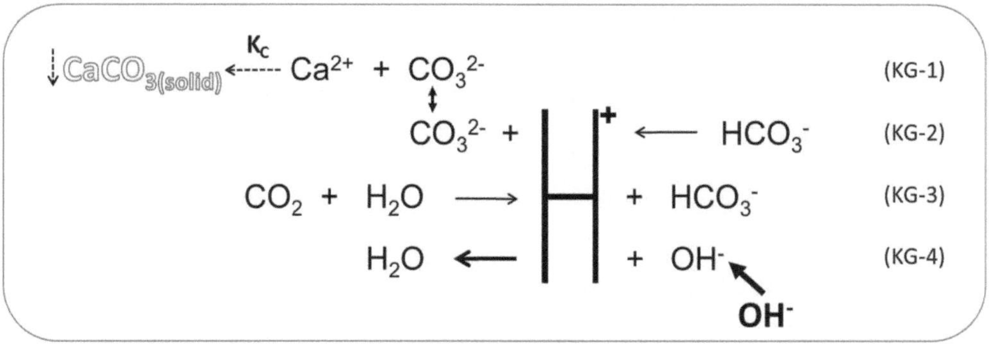

Figure 9.1-21: Deacidification with sodium hydroxide solution (NaOH)

- Lime can precipitate in case of overdosage of NaOH
- The water hardness remains unchanged
- The concentration of Na in the water increases

> **Dosing method / hydrated lime**

$$Ca(OH)_2 = Ca^{2+} + 2\ OH^-$$

The components effective for deacidification are the hydroxide ions and also the Ca^{2+} ion. The added hydroxide ions shift KG-4 (as with NaOH) to the left (Figure 9.1-22), consuming H^+ and increasing the pH. KG-2 and KG-3 partially compensate for the H^+ loss. The calcium ion increases the calcium concentration in KG-1.
KG-2 increases the carbonate concentration until calcite saturation is reached. If hydrated lime continues to be dosed, the solubility product is exceeded due to the further increasing carbonate concentration and the likewise increasing Ca concentration, and lime may precipitate.

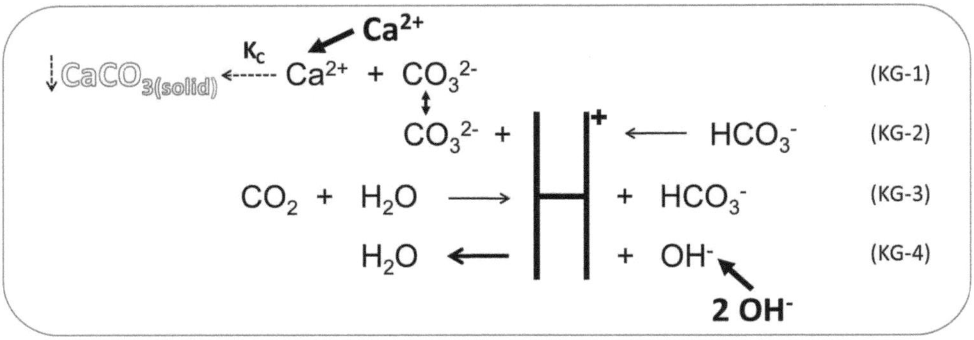

Figure 9.1-22: Deacidification with hydrated lime ($Ca(OH)_2$)

- Lime can precipitate in case of overdosage of $Ca(OH)_2$
- The water hardness increases

> **Dosing method / Sodium carbonate (soda)**

$$Na_2CO_3 \rightleftharpoons 2\ Na^+ + CO_3^{2-}$$

The components effective for deacidification are the carbonate ions.

The dosage of carbonate increases the concentration in KG-1 (Figure 9.1-23:), but also in KG-2, so that there is a shift of the reaction to the right and thus a consumption of H^+ (pH increase). KG-3 and KG-4 partially compensate for the H^+ loss.

If sodium carbonate continues to be dosed after the solubility product of $CaCO_3$ has been reached, the solubility will be exceeded and lime may precipitate.

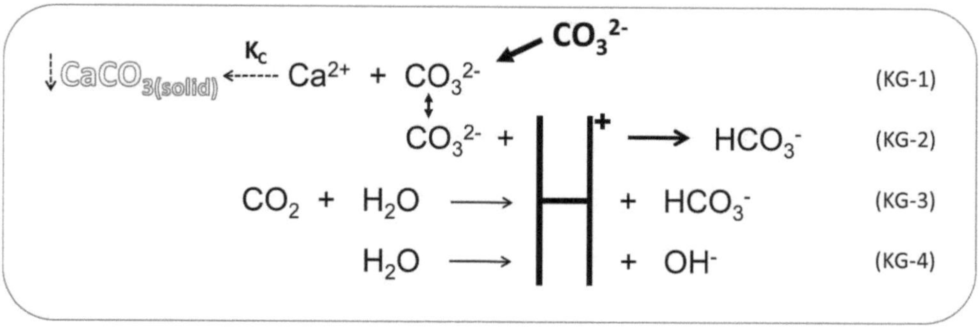

Figure 9.1-23: Deacidification with soda (Na_2CO_3)

- Lime can precipitate in case of overdosage of Na_2CO_3
- The water hardness remains unchanged
- The sodium concentration increases

➢ Filter method

<u>Limestone (calcite) = $CaCO_3$</u>

The component effective for deacidification is solid $CaCO_3$ as a filter material that can dissolve in water.

Carbonate is formed by dissolving lime and thus H^+ is consumed via KG-2, thus increasing the pH. The reaction proceeds until lime saturation (solubility product) is reached. Furthermore, since lime can no longer dissolve, supersaturation and consequent lime precipitation cannot occur.

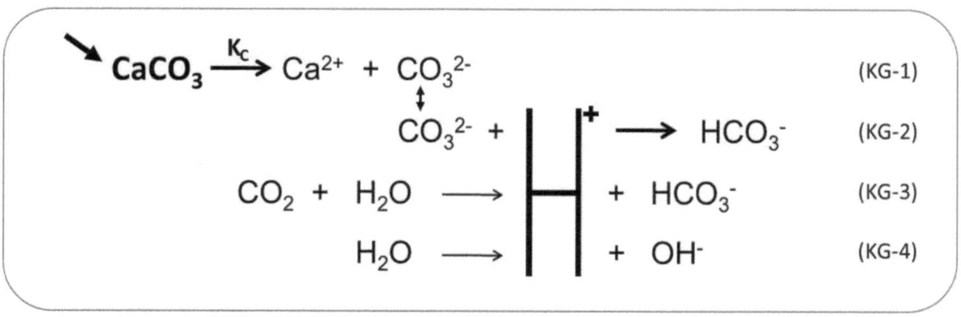

Figure 9.1-24: Deacidification with $CaCO_3$

- The solubility product <u>cannot</u> be exceeded !
- The water hardness increases

➢ Limestone (calcite) and additional hardening by dosing CO_2

In the case of particularly soft water, it may be useful to effect additional hardening by dosing CO_2 upstream of the filter stage.

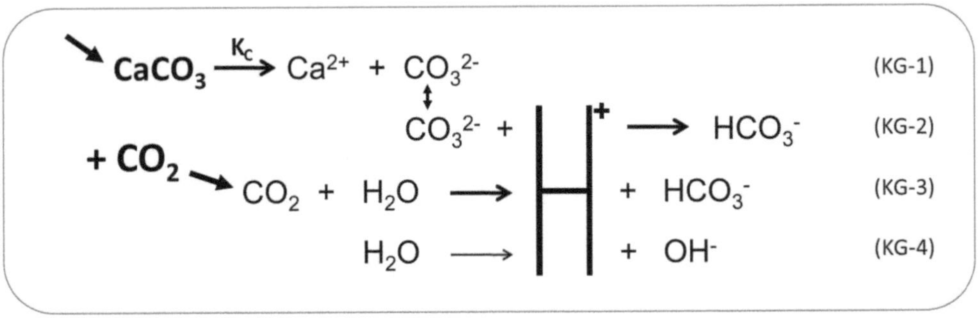

Figure 9.1-25: Deacidification with $CaCO_3$ and additional hardening by dosing CO_2

CO_2 dosing causes a shift in the KG-3, KG-4, and KG-2 reactions. Carbonate is consumed by KG-2, so more $CaCO_3$ must dissolve in KG-1 to reach the solubility limit. This increases the Ca^{2+} concentration and the hardness increases more than without the CO_2 dosage.

- The solubility product cannot be exceeded !
- The water hardness increases more

➢ Semi-fired dolomite (MgO / $CaCO_3$)

The semi-fired dolomite is used as a filter material. The calcium carbonate may dissolve in the water as described above. The magnesium oxide reacts with the water to form magnesium hydroxide. The OH^- ion is the second effective component for deacidification besides $CaCO_3$ (analogous to NaOH).

$$MgO + H_2O \rightleftharpoons Mg(OH)_2 \rightleftharpoons Mg^{2+} + 2\,OH^-$$

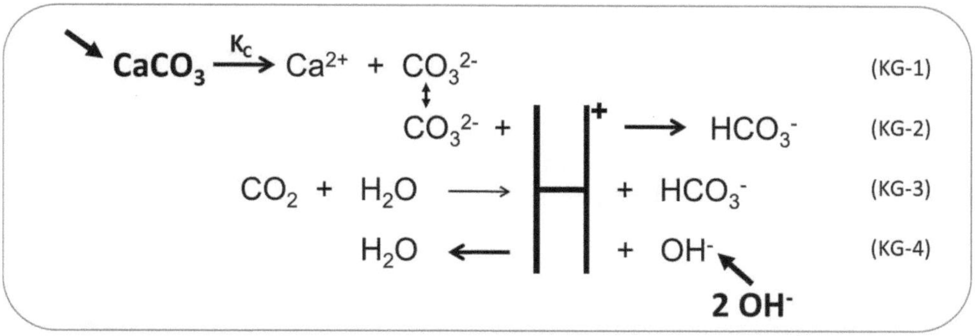

Figure 9.1-26: Deacidification with semi-fired dolomite (MgO/$CaCO_3$)

- Lime can precipitate if the filter speed is too low
- The water hardness increases

9.1.7 Final pH values of different deacidification methods

If different deacidification processes are applied to a water, the pH value after treatment will be different in each case. This means that the final pH value after deacidification depends on the deacidification process.

One can make the different pH values understandable with the reaction sequences shown.

The aim of deacidification is to reach the solubility limit for $CaCO_3$ and thus the lime-carbonic acid equilibrium.

For this purpose, either hydroxide ions, carbonate ions or calcium ions are added, calcite is dissolved or carbon dioxide is removed.

During deacidification with caustic soda, the pH is increased by the addition of hydroxide ions (KG-4) until sufficient carbonate has formed and the solubility limit for calcite is reached (Figure 9.1-21:)

If the same water is deacidified with hydrated lime (Figure 9.1-22:), the pH is also increased by adding the hydroxide ions. However, since the Ca^{2+} concentration is also increased at the same time (KG-1), less carbonate is needed to reach the solubility limit. This means that the pH does not have to be raised as much as with NaOH, the final pH will be lower.

The lowest final pH value is achieved during deacidification via lime filters ($CaCO_3$)(Table 9.1-5), since calcite dissolves directly here and Ca^{2+} and CO_3^{2-} ions are formed and the pH value is only increased by the secondary reaction (KG-2).

Example calculated values for the pH values for different deacidification processes are given in Table 9.1-5:.

Table 9.1-5: Final pH values for different deacidification methods

	pH value of calcite saturation	
Deacidifier	very soft water 0.4 mmol/L (2 °dH)	medium hard water 2.9 mmol/L (16 °dH)
$CaCO_3$	8.83	7.17
$Ca(OH)_2$	9.03	7.23
NaOH	9.16	7.29
- CO_2	--*	7.37

*Annotation: At 0.4 mmol/L (2°dH), deacidification by gas exchange is not possible (chapter 9.1.3).

9.2 Decarbonization / Softening

9.2.1 Delimitation of terms

The aim of decarbonization is to remove carbonic acid (forms of carbonic acid) from water or to reduce its content. Since softening often occurs simultaneously, or, since some softening processes are associated with decarbonization, the two terms are often used synonymously.

The terms used can be distinguished from each other by primarily referring to the **goal** of processing:

Decarbonization: Removal of carbonic acid (CO_2 , HCO_3^- , CO_3^{2-})

Softening: Removal of hardness components (Ca^{2+} and Mg^{2+})

Table 9.2-1: Differentiation between decarbonization and softening

Decarbonization	**Decarbonization = Softening**	**Softening**
Removal of carbonic acid CO_2 / HCO_3^- / CO_3^{2-}	Removal of hardness components **and** from carbonic acid	Removal of hardness components Ca^{2+} / Mg^{2+}
Ion exchanger (anion exchanger)	**Ion exchanger** (mixed bed exchanger)	**Ion exchanger** (cation exchanger)
	Reverse osmosis	**Nanofiltration**
	Precipitation of $CaCO_3$	
CO_2 removal through gas exchange		
without lime precipitation (Deacidification)	with lime precipitation	

In the precipitation processes and in reverse osmosis, softening and decarbonization occur simultaneously.

When using ion exchangers, both individual processes can be carried out separately, depending on whether an anion or a cation exchanger is used. When using mixed bed exchangers (full demineralization), decarbonization and softening occur simultaneously.

When removing CO_2 by gas exchange, one performs pure decarbonization (physical decarbonization = deacidification) as long as one does not exceed calcite saturation. As a rule, this process is used for the deacidification of water. If the calcite saturation is exceeded and lime precipitates, this process could be called physical softening. If the pH value of the water is lowered before the gas exchange by dosing acid, more CO_2 can be discharged from the water (acid decarbonization).

In membrane processes, reverse osmosis again involves both processes, but nanofiltration involves softening only.

Depending on whether the carbonic acid is removed from the water as far as is technically possible or whether it is only reduced by a specific amount, a distinction is made between the process variants **Full decarbonization** (full softening) and **partial decarbonization** (partial softening).

- Full decarbonization: decarbonization takes place completely (i.e. as far as procedurally possible)
- Partial decarbonization: decarbonization takes place only to a proportion

In the precipitation processes, full decarbonization reaches its process limit when there is no more calcium in the water to precipitate as carbonate, but residual concentrations of carbonic acid are still present in the water.

$$c(Ca^{2+}) < c(CO_3^{2-})$$

Full softening does not proceed completely in precipitation processes when there is no more carbonate in the water to precipitate, but residual concentrations of calcium are still present. Moreover, Mg^{2+} is not precipitated as carbonate, at high pH values at best as $Mg(OH)_2$, and remains partly as hardness in the water.

$$c(Ca^{2+}) > c(CO_3^{2-})$$

For treatment, it is possible to use the whole water flow, but to perform only partial decarbonization (**full flow partial decarbonization**). This treatment route is indicated for drinking water treatment when decarbonization is performed in combination with other treatment steps and the entire volume flow must be treated for the other treatment objective, e.g. flocculation, deferrization, demanganization.

If decarbonization is carried out alone as a treatment step, it can be useful in the treatment of drinking water to use only a partial stream of the water to be treated and to decarbonize it completely (**Partial stream full decarbonization**). Subsequently, the fully decarbonized partial stream is mixed again with the untreated main stream, whereby a partial decarbonization of the water is then achieved overall. This process route allows structurally smaller equipment to be used.

In the case of post-treatment of drinking water in private houses, partial flow full softening is always carried out during softening and then the correct degree of hardness is set by mixing with the untreated water.

For other treatment goals, it may be appropriate to fully decarbonize the entire water stream (**full stream full decarbonization**) (boiler feed water, fully desalinated water, pretreatment prior to membrane process).

In the case of process water in open cooling circuits, there is a concentration of water constituents. To prevent lime precipitation in the cooling system, these waters must be pretreated (decarbonization) or stabilized.

9.2.2 Precipitation process

Precipitation methods exceed the calcite solubility by raising the pH and separating the precipitated calcite.

The pH value is usually raised to 9-10. Raising the pH of water is achieved by dosing chemicals, in drinking water treatment almost always by adding calcium hydroxide ($Ca(OH)_2$) in the form of lime water or milk of lime, or more rarely with sodium hydroxide solution (NaOH). This pH range cannot be reached by gas exchange, but a partial increase of the pH value by gas exchange could be useful as a pre-treatment before the actual softening to save chemicals.

In precipitation methods must be taken into account that competitive precipitation may occur in this pH range. For example, depending on the conditions, magnesium hydroxide, iron hydroxide and iron carbonate may also precipitate.

The precipitation of iron results in a brown coloration of the precipitate product, which is often undesirable because of the utilization of the precipitated material.

As a rule, the pH is not set higher than 9-10 in drinking water treatment to avoid precipitation of magnesium hydroxide, since magnesium hydroxide has a negative effect on pellet formation (see below) and sludge quality.

Rapid decarbonization

In rapid decarbonization the water mixed with calcium hydroxide flows upstream through a fluidized bed of contact materials (quartz sand). The reaction vessel (reactor) is a vertically arranged cylindrical or conical vessel expanded at the top. The calcium carbonate accumulates here in the form of small balls, the so-called pellets.

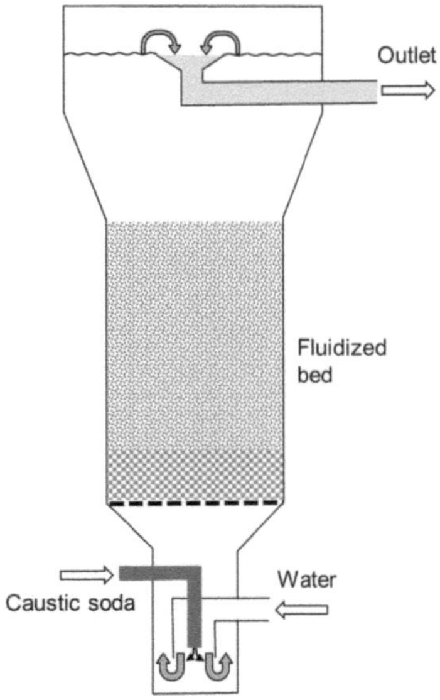

Figure 9.2-1: Fast decarbonizer, schematic

Figure 9.2-2: Pellets from a rapid decarbonization plant

Slow decarbonization

In the case of slow decarbonization, the actual precipitation process takes place in reaction basins in contact with material that has already been precipitated (sludge contact process). The product is a calcium carbonate slurry.

In the past, long residence times of the water to be treated were required to ensure sufficient reaction times. In addition, large settling ponds were required because calcium carbonate precipitates in fine crystals and requires long sedimentation times. However, through the development and optimization of increasingly efficient flocculation and sedimentation stages, total residence times of less than 20 min can be achieved today.

Figure 9.2-3: Slow decarbonization

9.2.3 Ion exchange process

When using ion exchange resins, a distinction must be made as to whether the treatment goal is decarbonization or softening.

Softening by ion exchange

For softening, cation exchangers are used, which exchange the hardness-forming cations calcium and magnesium for other cations. The type of ions exchanged depends on the regenerant for the exchangers. If the cation exchangers were regenerated with sodium chloride, sodium ions are released into the water during ion exchange; if they were regenerated with hydrochloric acid, H^+ ions are released.

The ion exchangers used for post-treatment in home softening are always in the Na form and are regenerated with NaCl.

<u>**Decarbonization by ion exchange**</u>

For pure decarbonization of a water one uses strongly basic anion exchangers.

9.2.4 Membrane process

Reverse osmosis results in complete desalination of the water. It is therefore a full softening and a partial decarbonization of the water, since gaseously dissolved CO_2 is not removed.

In nanofiltration, Ca^{2+} and Mg^{2+} ions are selectively retained by the membrane, so it is pure softening of water.

9.2.5 Effects of the procedures

Post-treatment of the water is necessary for all processes !

In **precipitation procedures**:

Filtration of suspended particles, possibly flocculation.

Since the water has usually been over-alkalized, the pH value must be stabilized by adding acid or admixing raw water.

In **ion exchange processes** and in **membrane processes**:

Since the softened water is supersaturated with CO_2, which must be removed

➔ physical deacidification.

10 Sample calculations

Calculations were **mostly** performed for the current (2022) atmospheric CO_2 concentration of **420 ppmv**. In some cases, however, other concentrations were used as a basis - please note and adapt to your own calculations!

10.1 Carbon content of the atmosphere

Example approximate calculation of the carbon content of the atmosphere based on the measured value for CO_2 in ppmv. (to chap. 2.4)

Calculation via the "homogeneous atmosphere" thought model [26].

For this purpose, it is assumed that the atmosphere above sea level stratifies homogeneously and does not dilute, producing a normal pressure of 1.01325 bar (101325 Pa) at sea level. The density of air at sea level of 1.293 kg/m³ at 0°C is assumed to be constant over the height of the air column.

Calculated for 420 ppmv CO_2 in the atmosphere.

Calculation of the height of the air column

Formula for the gravity pressure:

$$p = \rho \cdot g \cdot h \quad \longrightarrow \quad h = \frac{p}{\rho \cdot g}$$

$p = 101325\ \text{Pa}(=\text{N/m}^2)$
$\rho = 1.293\ \text{kg/m}^3$
$g = 9.81\ \text{N/kg}$

Atmospheric height $h = \frac{101325}{1{,}293 \cdot 9{,}81} \; \frac{\text{N} \cdot \text{m}^3 \cdot \text{kg}}{\text{m}^2 \cdot \text{kg} \cdot \text{N}} = 7945\ m = 7{,}945\ \text{km}$

Calculation of air volume V_{Air}

V_{air} = Earth surface A_{Earth} · Atmospheric height h

$A_{Earth} = 510 \cdot 10^6\ \text{km}^2$ [81]

$V_{air} = 510 \cdot 10^6\ \text{km}^2 \cdot 7.945\ \text{km} = 4.05 \cdot 10^9\ \text{km}^3 = 4.05 \cdot 10^{18}\ \text{m}^3$

Calculation volume CO_2 gas

Concentration ppmv = parts per million = factor 10^6

420 ppmv CO_2 (April 2022)

V_{CO2} = concentration ppmv · $V_{air}/10^6 = 1.70 \cdot 10^{15}\ m^3$

Calculation of the amount of $\mathbf{CO_2}$ n in mol at 0 °C

$V_{mol}(CO_2)$ = 22,41 L/mol = $22.41 \cdot 10^{-3}\ m^3/mol$

$n(CO_2) = V_{CO2}/V_{mol} = 1.70 \cdot 10^{15}\ m^3/22.41 \cdot 10^{-3}\ m^3/mol = 7.59 \cdot 10^{16}$ mol CO_2

Calculation $\mathbf{CO_2}$ mass m in gigatons

$M(CO_2)$= 44.0098 g/mol = 0.0440098 kg/mol

$m(CO_2) = n(CO_2) \cdot M(CO_2) = 3.31 \cdot 10^{15}$ kg = $3.31 \cdot 10^{12}$ t = 3.342 Gt CO_2

Calculation mass m of carbon **C** in Gt

M(C)=12.011 g/mol = 0.012011 kg/mol

$m(C) = n(CO_2) \cdot M(C) = 9.12 \cdot 10^{14}$ kg = $9.12 \cdot 10^{11}$ t = **912 Gt C**

10.2 CO_2 concentration in the air

(Calculations for chapter 3.1)

a) Molar volume of a gas [L/mol] at 10 °C ?

Data: n = number of moles = 1; R = 8.3144 J/(mol·K); T = absolute temperature in K; t = temperature in °C; p = 1 atm = 101,325 Pa (J/m³)

Solution:

General gas equation: $p \cdot V = n \cdot R \cdot T \quad \rightarrow \quad V = \frac{n \cdot R \cdot T}{p}$

or to Gay-Lussac:

$$V_{mol} = V_0 + V_0 \cdot \frac{t}{273.15} \quad \rightarrow \quad V_{mol} = 22.414 \cdot \frac{273.15+t}{273.15}$$

Table: Molar volume of a gas as a function of temperature (at 101,325 Pa)

Temp. t in °C	0	5	10	15	20	25
V_{mol} in L/mol	22.414	22.824	23.234	23.645	24.055	24.465

b) Molar concentration c_g and mass concentration β_g of CO_2 in the air at 10 °C (index g for gaseous)

Data: CO_2 content in air = 0.0420 vol%; $M(CO_2)$ = 44.009 g/mol

Concentration in air:

$$c_g = \frac{1}{V_{mol}\,[\text{L/mol}]} \cdot \frac{\text{vol\%}}{100} \quad [\text{mol/L}]$$

$$\beta_g = c_g \cdot M(CO_2) \qquad [\text{g/L}] \text{ or } [10^3\ \text{mg/L}]$$

Calculation for 10 °C:

$$c_g = \frac{1}{23.234\ \text{L/mol}} \cdot \frac{0.0420}{100} \quad [\text{mol/L}]$$

$$\boxed{c_g = 1.8077 \cdot 10^{-5}\ \text{mol/L}}$$

$$\beta_g = 1.8077 \cdot 10^{-5}\ \text{mol/L} \cdot 44.009\ \text{mg/}(10^{-3}\ \text{mol})$$

$$\boxed{\beta_g = 0.7956\ \text{mg/L}}$$

10.3 Solubility of CO_2 in water

(Calculations for chapter 4.1)

CO_2 dissolves in gaseous form in water according to the laws of HENRY.

$$c_W = K_H \cdot p \qquad \text{(Eq 10.3-1)}$$

$$p = \frac{\text{Vol.\%}}{100} \cdot p_{ges} \qquad \text{(Eq 10.3-2)}$$

c_W	Concentration of the gas in the water [mol/m3]
K_H	HENRY constant [mol/(m³·**bar**)]
p	Partial pressure [bar]
p_{ges}	Total pressure [bar]

Annotation: Henry's constant is still given in the literature with the unit bar (1 bar = 10^5 Pa ; 1 mbar = 1 hPa)

Solution of CO_2 at 10 °C and normal atmospheric pressure on contact water/air in water (saturation concentration):

Data: CO_2 content in air = **0.0420 vol%** $M(CO_2)$ = 44.009 g/mol

Table: Values for the HENRY constant (for calculation see chap. 7.1.1)

Temp. °C			0	5	10	15	20	25
K_H	Freshwater		76.895	63.418	53.077	45.026	38.674	33.600
mol/(m³·bar)	Salt water		62.303	52.006	43.969	37.626	32.571	28.503

Wanted: c_w and β_w (calculation with HENRY's law)

$$p = \frac{0.0420}{100} \cdot 1.01325 \text{ bar} = 0.000\,42557 \text{ bar}$$

c_W = 53.077 mol/(m³·bar) · 0.000 42557 bar

$\mathbf{c_w = 2.2588 \cdot 10^{-2}\ mol/m^3}$

$\beta_w = 2.2588 \cdot 10^{-2}$ mol/m³ · 44.009 g/mol = 0.99407 g/m³ = 0.99407 mg/L

β_w = **0.994 mg/L**

Temperature °C	CO_2 in water in mg/L	
	Freshwater I = 0	Seawater 1.9% Cl^-
0	1.44	1.17
5	1.19	0.974
10	**0.994**	0.824
15	0.843	0.705
20	0.724	0.610
25	0.629	0.534

! ! CO_2 in water
↓
takes from 0 - 20 °C
↓
down to half ! !

Air
↓↑

$$\mathbf{CO_2} + H_2O \rightleftharpoons H_2CO_3 \rightleftharpoons HCO_3^- + H^+ \rightleftharpoons CO_3^{2-} + 2\,H^+$$

↑
0.994 mg/L

10.4 Solution of CO_2 in leachate in equilibrium with the soil air (10 °C).

(Calculations for chapter 4.1)

Data: Soil air mostly 0.2 - 0.7 vol% CO_2
clay-rich, poorly drained soils up to over 5% [43]
$K_H(10°) = 53.077\ mol/(m^3 \cdot bar)$

Calculation for 0.2 vol% :

$c_W = K_H \cdot p$

$$p = \frac{0.2\%}{100} \cdot 1.01325\ bar = 0.0020265\ bar$$

$c_W = 53.077\ mol/(\ m^3 \cdot bar) \cdot 0.0020265\ bar = 0.10756\ mol/m^3$

$\beta_W = 0.10756\ mol/m^3 \cdot 44.009\ g/mol = 4.7336\ g/m^3$

Tabelle 10.4-1: CO_2 contents in water as a function of the partial pressure of CO_2 in the gas phase.

CO_2 (g) vol%	Partial pressure bar	CO_2 in water mol/m³	CO_2 in water mg/L
0.2	0.0020265	0.10756	4.734
0.7	0.0070928	0.37646	16.57
1	0.0101325	0.53780	23.67
2	0.0202650	1.0756	47.34
3	0.0303975	1.6134	71.00
5	0.0506625	2.6890	118.34

In carbonic acid analysis, the CO_2 dissolved in water in gaseous form is determined with the base capacity $K_{B8.2}$. The measured value of the base capacity would therefore correspond approximately to the 3rd column of the table.

Depending on the amount of CO_2 dissolved in the water, the pH value of the water changes (➔ Example 10.5).

10.5 pH value of a CO_2 solution

(Calculations for chapter 4.1)

pH value of a CO_2 solution (at 10 °C) (420 ppmv)

Data: K_1 (10 °C) = $3.4259 \cdot 10^{-7}$ mol/L

$a(CO_2) = c_w = 2.2588 \cdot 10^{-5}$ mol/L (see above example 0)

a) Rainwater / aqua dest. - in balance with the air

$$CO_2 + H_2O \leftrightarrows HCO_3^- + H^+$$

it follows: $a(HCO_3^-) = a(H^+)$

Approximations: - Self-dissociation of the water is neglected !
- Strongly diluted solution, I very small → a(i) = c(i) !

insert in : $\frac{a(H^+)\cdot a(HCO_3^-)}{a(CO_2)} = K_1$

$\rightarrow \frac{a(H^+)^2}{a(CO_2)} = K_1 \rightarrow a(H^+)^2 = K_1 \cdot a(CO_2)$

$a(H^+)^2 = 3.4259 \cdot 10^{-7}$ mol/L $\cdot$ $2.2588 \cdot 10^{-5}$ mol/L

$a(H^+)^2 = 7.7383 \cdot 10^{-12}$ (mol/L)^2

$a(H^+) = 2.7818 \cdot 10^{-6}$ mol/L

$-\log(a(H^+))$ = pH = 5.556

Table 10.5-1: Calculated pH values for rainwater

Year	CO_2 vol% in air	water pH value
ca.1750	0.0280	5.644
1960	0.0317	5.617
1970	0.0326	5.611
1980	0.0339	5.602
1990	0.0354	5.593
2000	0.0370	5.583
2010	0.0390	5.572
2020	0.0414	5.559

b) Soil air with 0.7 % by volume CO_2

Data: $a(CO_2) = c_w = 3.7645 \cdot 10^{-4}$ mol/L (see above example 10.4)

$a(H^+)^2 = 3.4259 \cdot 10^{-7}$ mol/L $\cdot$ $3.7646 \cdot 10^{-4}$ mol/L

$a(H^+)^2 = 1.2897 \cdot 10^{-10}$ (mol/L)^2

$a(H^+) = 1.1357 \cdot 10^{-5}$ mol/L

$-\log(H^+)$ = pH = 4.94

The pH value of 4.94 is only theoretical, since groundwaters generally have pH values above 6. In the soil area, balancing reactions occur (e.g. ion exchange processes, dissolution of lime).

Table 10.5-2: Theoretical pH values in equilibrium with the soil air

CO_2 vol%	pH value in water
0.0420	5.56
0.2	5.22
0.7	4.94
1	4.87
2	4.72
3	4.63
5	4.52

For comparison, the value for atmospheric air (0.0420 vol%) is included.

10.6 Supplementary consideration of the lime-carbonic acid system

(To chapter 4.7)

It can be shown by converting the mass action equations that for the same CO_2 concentration, the HCO_3^- concentration behaves inversely proportional to the H^+ concentration, while the CO_3^{2-} concentration depends inversely proportional to the square of the H^+ concentration.
This leads to different slopes of the graphs in Figure 6.1-3.

$$\frac{a(H^+)\cdot a(HCO_3^-)}{a(CO_2)} = K_1 \qquad (1)$$

$$\frac{a(CO_3^{2-})\cdot a(H^+)}{a(HCO_3^-)} = K_2 \qquad (2)$$

↓

$$a(HCO_3^-) = \frac{a(CO_2)\cdot K_1}{a(H^+)} \qquad (3)$$

$$a(CO_3^{2-}) = \frac{a(HCO_3^-)\cdot K_2}{a(H^+)} \qquad (4)$$

↓

(3) in (4)

$$a(CO_3^{2-}) = \frac{a(CO_2)\cdot K_1\cdot K_2}{[a(H^+)]^2} \qquad (5)$$

10.7 Calculation of the carbonic acid content

(Calculations for chapter 4.5.3)

(The numbering of the reaction equations refers to this example only)

The **carbonic acid sum** c(C) is **constant** and independent of temperature and ionic strength in a **closed** system, i.e. without exchange with the atmosphere (no precipitation or solution !).

$$c(C) = c(CO_2) + c(HCO_3^-) + c(CO_3^{2-}) \qquad \textbf{(1)}$$

However, the proportions in which the different forms of carbonic acid are present depend on

- pH value
- Temperature
- Ionic strength

For a given temperature (here: 10 °C) and ionic strength (here: I=0), the proportions $x(CO_2)$, $x(HCO_3^-)$ and $x(CO_3^{2-})$ of the carbonic acid forms can be calculated from the mass balance via the mass action equations as a function of pH.

Shares of the carbonic acid forms in the carbonic acid sum c(C)

$$c(CO_2) = x(CO_2) \cdot c(C) \quad \textbf{(2)} \quad \rightarrow \quad x(CO_2) = \frac{c(CO_2)}{c(C)} \quad \textbf{(3)}$$

$$c(HCO_3^-) = x(HCO_3^-) \cdot c(C) \quad \textbf{(4)} \quad \rightarrow \quad x(HCO_3^-) = \frac{c(HCO_3^-)}{c(C)} \quad \textbf{(5)}$$

$$c(CO_3^{2-}) = x(CO_3^{2-}) \cdot c(C) \quad \textbf{(6)} \quad \rightarrow \quad x(CO_3^{2-}) = \frac{c(CO_3^{2-})}{c(C)} \quad \textbf{(7)}$$

The basis of the calculations is the molar balance:

$$c(C) = c(CO_2) + c(HCO_3^-) + c(CO_3^{2-}) \qquad \textbf{(1)}$$

Procedure:	↓ Share $x(CO_2)$ **searched**	↓ ↓ replace with CO_2 via the corresponding Mass action equations

1. Equilibrium constant

$$\frac{a(H^+) \cdot c(HCO_3^-) \cdot f}{c(CO_2)} = K_1 \quad \textbf{(8)} \quad \rightarrow \quad c(HCO_3^-) = \frac{c(CO_2) \cdot K_1}{a(H^+) \cdot f} \quad \textbf{(9)}$$

2. Equilibrium constant

$$\frac{c(CO_3^{2-}) \cdot f^4 \cdot a(H^+)}{c(HCO_3^-) \cdot f} = K_2 \quad \textbf{(10)} \quad \rightarrow \quad c(CO_3^{2-}) = \frac{c(HCO_3^-) \cdot K_2}{a(H^+) \cdot f^3} \quad \textbf{(11)}$$

(9) in (11) $$c(CO_3^{2-}) = \frac{\frac{c(CO_2)\cdot K_1}{a(H^+)\cdot f}\cdot K_2}{a(H^+)\cdot f^3} \rightarrow \quad c(CO_3^{2-}) = \frac{c(CO_2)\cdot K_1\cdot K_2}{a(H^+)^2\cdot f^4} \quad (12)$$

Insert into the molar balance:

(9) and **(12)** in **(1)** : $$c(C) = c(CO_2) + \frac{c(CO_2)\cdot K_1}{a(H^+)\cdot f} + \frac{c(CO_2)\cdot K_1\cdot K_2}{a(H^+)^2\cdot f^4} \quad (13)$$

transform: $$c(C) = c(CO_2)\cdot\left[1 + \frac{K_1}{a(H^+)\cdot f} + \frac{K_1\cdot K_2}{a(H^+)^2\cdot f^4}\right] \quad (14)$$

(3): $$\frac{c(CO_2)}{c(C)} = \mathbf{x(CO_2)} = \frac{1}{1+\frac{K_1}{a(H^+)\cdot f}+\frac{K_1\cdot K_2}{a(H^+)^2\cdot f^4}} \quad (15)$$

The same procedure is followed for the other factors:

(2) in (9) $$c(HCO_3^-) = \frac{x(CO_2)\cdot c(C)\cdot K_1}{a(H^+)\cdot f} \rightarrow c(HCO_3^-) = \frac{x(CO_2)\cdot K_1}{a(H^+)\cdot f}\cdot c(C) \quad (16)$$

(5): $$\frac{c(HCO_3^-)}{c(C)} = \mathbf{x(HCO_3^-)} = \frac{x(CO_2)\cdot K_1}{a(H^+)\cdot f} \quad (17)$$

(4) in (11) $$c(CO_3^{2-}) = \frac{x(HCO_3^-)\cdot c(C)\cdot K_2}{a(H^+)} \rightarrow c(CO_3^{2-}) = \frac{x(HCO_3^-)\cdot K_2}{a(H^+)}\cdot c(C) \quad (18)$$

(7): $$\frac{c(CO_3^{2-})}{c(C)} = x(CO_3^{2-}) = \frac{x(HCO_3^-)\cdot K_2}{a(H^+)\cdot f^3} \quad (19)$$

or with **(17)** in **(19)**: $$\mathbf{x(CO_3^{2-})} = \frac{x(CO_2)\cdot K_1\cdot K_2}{a(H^+)^2\cdot f^4} \quad (20)$$

The factors $x(CO_2)$, $x(HCO_3^-)$ and $x(CO_3^{2-})$ are only dependent on the pH value at a given temperature and ionic strength (here I = 0) !

$K_1(10°C) = 3.426 \cdot 10^{-7}$ mol/L
$K_2(10°C) = 3.249 \cdot 10^{-11}$ mol/L

Table 10.7-1: Relative proportions of the carbonic acid forms as a function of pH (10 °C) (I=0) (Further values in the table appendix 12.3)

pH	$x(CO_2)$	$x(HCO_3^-)$	$x(CO_3^{2-})$
6.0	0.745	0.255	0.00000829
7.0	0.226	0.774	0.000251
8.0	0.0283	0.969	0.00315
9.0	0.00282	0.966	0.0314
10.0	0.000220	0.755	0.245

For graphical representation see chapter 4.5.

10.7.1 Calculation of carbonic acid contents with complex formation

(The numbering of the reaction equations refers to this example only)

The aim is to show the influence of complex formation on the distribution of carbonic acid contents.
Calculation for a closed system analogous to chap. 10.7 .

Basis is the carbonic acid sum c(**tC**) with complexes:

$$c(\mathbf{tC}) = c(CO_2) + c(tHCO_3^-) + c(tCO_3^{2-}) \quad \textbf{(1)}$$

$$c(tC) = c(CO_2) + c(HCO_3^-) \cdot F_{C1} + c(CO_3^{2-}) \cdot F_{C2} \quad \textbf{(2)}$$

$$F_{C1} = 1 + c(Ca^{2+}) \cdot f^4 \cdot K_3 + c(Mg^{2+}) \cdot f^4 \cdot K_{10} \quad \textbf{(3)}$$

$$F_{C2} = 1 + c(Ca^{2+}) \cdot f^8 \cdot K_4 + c(Mg^{2+}) \cdot f^8 \cdot K_{11} \quad \textbf{(4)}$$

Temperature (10 °C) Ionic strength (here: I=0)

1. Equilibrium constant

$$\frac{a(H^+) \cdot c(HCO_3^-) \cdot f}{c(CO_2)} = K_1 \quad \rightarrow \quad c(HCO_3^-) = \frac{c(CO_2) \cdot K_1}{a(H^+) \cdot f} \quad \textbf{(5)}$$

2. Equilibrium constant

$$\frac{c(CO_3^{2-}) \cdot f^4 \cdot a(H^+)}{c(HCO_3^-) \cdot f} = K_2 \quad \rightarrow \quad c(CO_3^{2-}) = \frac{c(HCO_3^-) \cdot K_2}{a(H^+) \cdot f^3} \quad \textbf{(6)}$$

(3) in **(4)**
$$c(CO_3^{2-}) = \frac{\frac{c(CO_2) \cdot K_1}{a(H^+) \cdot f} \cdot K_2}{a(H^+) \cdot f^3} \rightarrow \quad c(CO_3^{2-}) = \frac{c(CO_2) \cdot K_1 \cdot K_2}{a(H^+)^2 \cdot f^4} \quad \textbf{(7)}$$

Insert into the molar balance:

$$c(tC) = c(CO_2) + c(HCO_3^-) \cdot F_{C1} + c(CO_3^{2-}) \cdot F_{C2} \quad \textbf{(2)}$$

(5)+(7) in **(2)**
$$c(tC) = c(CO_2) + \frac{c(CO_2) \cdot K_1}{a(H^+) \cdot f} \cdot F_{C1} + \frac{c(CO_2) \cdot K_1 \cdot K_2}{a(H^+)^2 \cdot f^4} \cdot F_{C2} \quad \textbf{(8)}$$

$$c(tC) = c(CO_2) \cdot \left[1 + \frac{K_1}{a(H^+) \cdot f} \cdot F_{C1} + \frac{K_1 \cdot K_2}{a(H^+)^2 \cdot f^4} \cdot F_{C2}\right] \quad \textbf{(9)}$$

$$c(CO_2) = \frac{c(tC)}{\left[1 + \frac{K_1}{a(H^+) \cdot f} \cdot F_{C1} + \frac{K_1 \cdot K_2}{a(H^+)^2 \cdot f^4} \cdot F_{C2}\right]} \quad \textbf{(10)}$$

Shares of the carbonic acid forms $x(CO_2)$, $x(HCO_3^-)$ and $x(CO_3^{2-})$ in the carbonic acid sum c(tC):

$$x(CO_2) = \frac{c(CO_2)}{c(tC)} = \frac{1}{\left[1+\frac{K_1}{a(H^+)\cdot f}\cdot F_{C1}+\frac{K_1\cdot K_2}{a(H^+)^2\cdot f^4}\cdot F_{C2}\right]} \quad (11)$$

The same procedure is followed for the other factors:

$$\mathbf{x(HCO_3^-)} = \frac{c(HCO_3^-)}{c(tC)} = \frac{x(CO_2)\cdot K_1}{a(H^+)\cdot f} \quad (12)$$

$$\mathbf{x(HCO_3^-)} = \frac{K_1}{\left[1+\frac{K_1}{a(H^+)\cdot f}\cdot F_{C1}+\frac{K_1\cdot K_2}{a(H^+)^2\cdot f^4}\cdot F_{C2}\right]\cdot a(H^+)\cdot f} \quad (13)$$

$$\mathbf{x(HCO_3^-)} = \frac{K_1}{a(H^+)\cdot f+K_1\cdot F_{C1}+\frac{K_1\cdot K_2}{a(H^+)\cdot f^3}\cdot F_{C2}} \quad (14)$$

$$\mathbf{x(HCO_3^-)} = \frac{1}{\frac{a(H^+)\cdot f}{K_1}+\cdot F_{C1}+\frac{\cdot K_2}{a(H^+)\cdot f^3}\cdot F_{C2}} \quad (15)$$

$$\mathbf{x(CO_3^{2-})} = \frac{c(CO_3^{2-})}{c(tC)} = \frac{x(HCO_3^-)\cdot K_2}{a(H^+)\cdot f^3} = \frac{x(CO_2)\cdot K_1\cdot K_2}{a(H^+)^2\cdot f^4} \quad (16)$$

$$\mathbf{x(CO_3^{2-})} = \frac{K_1\cdot K_2}{\left[1+\frac{K_1}{a(H^+)\cdot f}\cdot F_{C1}+\frac{K_1\cdot K_2}{a(H^+)^2\cdot f^4}\cdot F_{C2}\right]\cdot a(H^+)^2\cdot f^4} \quad (17)$$

$$\mathbf{x(CO_3^{2-})} = \frac{K_1\cdot K_2}{a(H^+)^2\cdot f^4+a(H^+)\cdot K_1\cdot f^3\cdot F_{C1}+K_1\cdot K_2\cdot F_{C2}} \quad (18)$$

The factors $x(CO_2)$, $x(HCO_3^-)$ and $x(CO_3^{2-})$ are only dependent on the pH value at a given temperature and ionic strength (here I = 0) !

Table: Relative proportions of carbonic acid forms as a function of pH (10 °C) (I=9.7 mmol/L, hardness 2,7 mmol/L (15°dH)) taking complex formation into account

(**without / with** complexes)

pH	$x(CO_2)$	$x(HCO_3^-)$	$x(CO_3^{2-})$
6.0	0.696 / 0.694	0.304 / 0.303	0.0000 / 0.0000
7.0	0.186 / 0.185	0.813 / 0.806	0.0005 / 0.0005
8.0	0.0222 / 0.0222	0.971 / 0.959	0.0065 / 0.0064
9.0	0.0021 / 0.0021	0.935 / 0.903	0.0628 / 0.0606
10.0	0.0001 / 0.0001	0.598 / 0.510	0.402 / 0.343
11.0	0.0000 / 0.0000	0.130 / 0.095	0.871 / 0.0.640

For graphical representation see chapter 4.5.3.

10.8 Thought model for carbonic acid in water: Open system

(Calculations for chapter 6.1)

Open system - always gas exchange with air with equilibrium adjustment

↓↑

$$CO_2 + H_2O \rightleftharpoons HCO_3^- + H^+ \rightleftharpoons CO_3^{2-} + 2\,H^+$$

$c_W = c(CO_2) =$ **constant**

The proportions of the carbonic acid forms can be calculated (calculation example 10.7) and are thus known. Therefore, c(C) can be calculated with $c(CO_2)$ from the available data.

$$c(CO_2) = x(CO_2) \cdot c(C) \quad \rightarrow \quad c(C) = \frac{c(CO_2)}{x(CO_2)}$$

The concentrations $c(HCO_3^-)$ and $c(CO_3^{2-})$ can then be calculated from c(C) via $x(HCO_3^-)$ and $x(CO_3^{2-})$:

$$c(HCO_3^-) = x(HCO_3^-) \cdot c(C)$$

$$c(CO_3^{2-}) = x(CO_3^{2-}) \cdot c(C)$$

a) Air with **0.0400%** CO_2 by volume (10°C):
$c_W = c(CO_2) = \text{constant} = 0.021512 \cdot 10^{-2}$ mol/m³

Example: Calculation for pH 7.0 (10 °C) (concentrations in mol/m³)

pH	$x(CO_2)$	$x(HCO_3^-)$	$x(CO_3^{2-})$
7.0	0.226	0.774	$0.251 \cdot 10^{-3}$

$$c(C) = \frac{c(CO_2)}{x(CO_2)} = \frac{0.021512 \cdot 10^{-2}}{0.226} = 0.0952$$

$$c(HCO_3^-) = x(HCO_3^-) \cdot c(C) = 0.774 \cdot 0.0952 = 0.07368$$

$$c(CO_3^{2-}) = x(CO_3^{2-}) \cdot c(C) = 0.251 \cdot 10^{-3} \cdot 0.0952 = 0.23895 \cdot 10^{-4}$$

Table 10.8-1: Results Example a), concentrations in mol/m³

		CO_2	HCO_3^-	CO_3^{2-}	C
pH 5.6	Share	0.880	0.120	$0.155 \cdot 10^{-5}$	1
	Concentration	0.0215	0.00293	$0.380 \cdot 10^{-7}$	0.0244
pH 7.0	Share	0.226	0.774	$0.251 \cdot 10^{-3}$	1
	Concentration	0.0215	0.0737	$0.239 \cdot 10^{-4}$	0.0952
pH 8.0	Share	0.0283	0.969	0.00315	1
	Concentration	0.0215	0.737	0.00239	0.761
pH 9.0	Share	0.00282	0.966	0.0314	1
	Concentration	0.0215	7.37	0.239	7.63

b) Soil air with 2.0% CO_2 by volume (10 °C): $c_W = c(CO_2)$ = constant = 1.076 mol/m³

Table 10.8-2: Results Example b), concentrations in mol/m³

		CO_2	HCO_3^-	CO_3^{2-}	C
pH 6.0	Share	0.745	0.255	$0.829 \cdot 10^{-5}$	1
	Concentration	1.076	0.369	$0.120 \cdot 10^{-4}$	1.45
pH 7.0	Share	0.226	0.774	$0.251 \cdot 10^{-3}$	1
	Concentration	1.076	3.69	0.00120	4.76
pH 8.0	Share	0.0283	0.969	0.00315	1
	Concentration	1.076	36.9	0.120	38.1
		$\approx K_{B8.2}$!	$\approx K_{S4,3}$!		**TIC !**

Calculation of the boundary line Ca^{2+} / $CaCO_3$

Calculation of the Ca^{2+} concentration when the solubility product is reached: Limiting concentration (K_C ($CaCO_3$,10 °C) = $3.927 \cdot 10^{-3}$ (mol/m³)²)

$$K_C = a(Ca^{2+}) \cdot a(CO_3^{2-}) \qquad a(Ca^{2+}) = \frac{K_C}{a(CO_3^{2-})}$$

Water a)

pH 7 : $a(Ca^{2+}) = 3.927 \cdot 10^{-3} / 0{,}239 \cdot 10^{-4} = 1.64 \cdot 10^{+2}$

pH 8 : $a(Ca^{2+}) = 3.927 \cdot 10^{-3} / 0{,}239 \cdot 10^{-2} = 1.64$

pH 9 : $a(Ca^{2+}) = 3.927 \cdot 10^{-3} / 0.239 = 1.64 \cdot 10^{-2}$

Water b) pH 6 : $a(Ca^{2+}) = 3.927 \cdot 10^{-3} / 0{,}120 \cdot 10^{-4} = 3.28 \cdot 10^{+2}$
pH 7 : $a(Ca^{2+}) = 3.927 \cdot 10^{-3} / 0{,}120 \cdot 10^{-2} = 3.28$
pH 8 : $a(Ca^{2+}) = 3.927 \cdot 10^{-3} / 0{,}120 = 3.28 \cdot 10^{-2}$

Water hardness ranges

Upper limit for drinking water : 5.4 mmol/L (30 °dH)
Lower limit: Soft reservoir water for example: 0.5 mmol/L (3 °dH)
Conversion factor $mol/m^3 \rightarrow$ hardness °dH = 5.6
assuming: $a(Ca^{2+})$ = 80% of the hardness → $(30\ °dH \cdot 0.8)/5.6 = 4.28\ mol/m^3\ Ca^{2+}$
$(3\ °dH \cdot 0.8)/5.6 = 0.429\ mol/m^3\ Ca^{2+}$

10.9 Calculation of the carbonic acid shapes from the TIC

(Calculations for chapter 7.2.1)

From the TIC, the concentrations of the carbonic acid forms can be calculated via the relative proportions of the carbonic acid forms (dependent on pH!).

$c(C) = TIC \quad \rightarrow \quad c(CO_2) = x(CO_2) \cdot c(C)$

$c(HCO_3^-) = x(HCO_3^-) \cdot c(C)$

$c(CO_3^{2-}) = x(CO_3^{2-}) \cdot c(C)$

Data:

$M(C)$	= 12.011	g/mol	$M(HCO_3^-)$	= 61.017	g/mol
$M(CO_2)$	= 44.010	g/mol	$M(CO_3^{2-})$	= 60.009	g/mol

<u>Example :</u> TIC = 3.6 mmol/L at pH = 7.0

a) Molar concentration of inorganic carbon (carbonic acid sum):

$c(C) = TIC = 3.6\ mmol/L$

b) Mass concentration of inorganic carbon:

$\beta(C) = c(C) \cdot M(C) = 3.6\ mmol/L \cdot 12.011\ mg/mmol = 43.2\ mg/L\ C$

c) Concentrations of the respective forms of carbonic acid:

The proportions can be taken from Table 12.3-1::

Shares at pH 7.0	
	$x(CO_2) = 0.226$
	$x(HCO_3^-) = 0.774$
	$x(CO_3^{2-}) = 0.251 \cdot 10^{-3}$

$c(CO_2)$	$= x(CO_2) \cdot c(C)$	$= 0.226$	$\cdot 3.6$ mmol/L	$= 0.814$	mmol/L
$c(HCO_3^-)$	$= x(HCO_3^-) \cdot c(C)$	$= 0.774$	$\cdot 3.6$ mmol/L	$= 2.78$	mmol/L
$c(CO_3^{2-})$	$= x(CO_3^{2-}) \cdot c(C)$	$= 0.251 \cdot 10^{-3}$	$\cdot 3.6$ mmol/L	$= 0.904 \cdot 10^{-3}$	mmol/L

$\beta(CO_2)$	$= c(CO_2) \cdot M(CO_2)$	$= 0.814$ mmol/L	$\cdot 44.010$	mg/mmol	$= 35.8$	mg/L
$\beta(HCO_3^-)$	$= c(HCO_3^-) \cdot M(HCO_3^-)$	$= 2.78$ mmol/L	$\cdot 61.017$	mg/mmol	$= 170.0$	mg/L
$\beta(CO_3^{2-})$	$= c(CO_3^{2-}) \cdot M(CO_3^{2-})$	$= 0.904 \cdot 10^{-3}$ mmol/L	$\cdot 60.009$	mg/mmol	$= 0.05$	mg/L

10.10 Acid capacity of pure water

(Calculations for chapter 7.2.2.1)

If the water sample is ultrapure water (I = 0), only the dissociation equilibrium of the water must be shifted by the titration:

$$H_2O \rightleftharpoons H^+ + OH^-$$

The dilution effect that occurs with all titrations can be neglected by adding only small volumes if the molarities of the titrant are chosen correctly.

Example acid capacity: Addition of H^+ :

In a water with pH 7, the acid capacity 4.3 is determined. This changes the pH from 7.0 to 4.3 by adding acid. The concentration of H^+ ions (25 °C: $K_W = 10^{-14}$ $(mol/L)^2$) is thus increased from 10^{-7} mol/L to $10^{-4.3}$ mol/L:

pH	= 7.0	pH	= 4.3
$a(H^+)$	= 10^{-7} mol/L	$a(H^+)$	= $10^{-4.3}$ mol/L
$a(OH^-)$	= 10^{-7} mol/L	$a(OH^-)$	= $10^{-9.7}$ mol/L

At the end there are : $a(H^+) = 10^{-4.3}$ and $a(OH^-) = 10^{-9.7}$ mol/L

Part of the added H^+ ions is consumed by the neutralization of OH^- ions . The other part increases the H^+ concentration in the water to the desired pH.

Neutralization OH^- ions $\quad 10^{-7} - 10^{-9.7} = H_{consumption} = 9.98 \cdot 10^{-8}$ mol/L

Decrease pH value $\quad 10^{-4.3} - 10^{-7} = H_{change} = 5001.87 \cdot 10^{-8}$ mol/L +

$H_{Titration\ consumption} = 5011.85 \cdot 10^{-8}$ mol/L

Thus, a total of $5.01185 \cdot 10^{-5}$ mol/L of H^+ ions are consumed.

The consumption for neutralization ($9.98 \cdot 10^{-8}$ mol/L) can be neglected because of the low value.

Thus, the H^+ consumption for shifting the dissociation equilibrium of water is about $0.05 \cdot 10^{-3}$ mol/L at magnitudes of normal acid capacities in drinking waters from $1 \cdot 10^{-3}$ to $5 \cdot 10^{-3}$ mol/L.

10.11 Saturation index

(Calculation for chapter 8.5. This variant can be calculated with a calculator)

The calculation of the saturation index is based on analytical data taking into account the ionic strength.

$$SI = \lg\left(\frac{c(Ca^{2+})\cdot c(HCO_3^-)\cdot K_2}{K_C\cdot a(H^+)}\cdot f^5\right)$$

Since the activity of the H^+ ions is recorded directly with the pH electrode, $a(H^+)$ can be derived directly from the pH value and does not have to be taken into account when calculating the activity coefficient!

Required values: Temperature T in K (for calculation of the constants)
pH
$c(Ca^{2+})$ in **mol/L**
$K_{S4.3}$ in **mol/L**
el. conductivity κ_{25} in **mS/m**
t_e
t_b
t_{pH}

Example - Analysis		
t_e	°C	10.0
t_b	°C	10.0
pH		7.10
t_{pH}	°C	18
LF (κ_{25})	mS/m	65
$K_{S4.3}$	mmol/L	2.8
Ca^{2+}	mmol/L	2.5

K_2 (10°C) in mol/L	$3.249\cdot10^{-11}$
K_C (10°C) $(mol/L)^2$	$3.927\cdot10^{-9}$

approximated: $c(HCO_3^-) \approx K_{S4.3}$

Temperature correction of the pH value:

If the temperature (t_{pH}) deviates from the evaluation temperature (t_b) when measuring the pH value, the pH value (pH) measured at this temperature must be converted to the evaluation temperature (pH_{tb}).

$pH_{tb} = pH + 0.011 \cdot (t_{pH} - t_b)$

$pH_{tb} = 7.10 + 0.011 \cdot (18 - 10) = 7.19$

Calculation of the ionic strength I approximately from the conductivity:

$$I = \frac{\kappa_{25}}{6200} \text{ [mol/L]} \qquad (\kappa_{25} \text{ in mS/m})]$$

➔ $I = 65/6200 = 0.01048$ mol/L

Calculation of the activity coefficient f:

$$\lg(f) = -0.5 \cdot \frac{\sqrt{I}}{1+1.4 \cdot \sqrt{I}}$$

➔ $$\lg(f) = -0.5 \cdot \frac{\sqrt{0.01048}}{1+1.4 \cdot \sqrt{0.01048}} = -0.0448$$

$f = 10^{\lg(f)}$ ➔ $f = 10^{-0,0448} = 0.902$

Calculation of hydrogen ion activity

$a(H^+) = 10^{-pH}$ ➔ $a(H^+) = 10^{-7.19} = 6.49 \cdot 10^{-8}$

Insert into the equation

$$SI = \lg\left(\frac{c(Ca^{2+}) \cdot c(HCO_3^-) \cdot K_2}{K_C \cdot a(H^+)} \cdot f^5\right)$$ Units in **mol/L**

$$SI = \lg\left(\frac{2.5 \cdot 10^{-3} \cdot 2.8 \cdot 10^{-3} \cdot 3.249 \cdot 10^{-11}}{3.927 \cdot 10^{-9} \cdot 6.49 \cdot 10^{-8}} \cdot 0.902^5\right)$$

$SI = \lg(0.5332)$

<u>SI = - 0.273</u>

SI < 0 : The water is calcite soluble

11 Chemical Basics in Brief

11.1 Basic terms and concentration information

Amount of substance n

The amount of substance n is a basic quantity (SI unit)
The unit of the amount of substance is the mole [mol]
1 mole is the amount of substance n consisting of N_A particles
(number of atoms in one mole of carbon ^{12}C: $N_A = 6.022 \cdot 10^{23}\ mol^{-1}$
(Avogadro constant or LOSCHMIDT number))

Molar mass M(i) (molecular weight) of substance (i)

Link between the quantity of substance n and the associated mass m

$$M(i) = \frac{m(i)}{n(i)} \quad \frac{g}{mol} \qquad \text{or also} \qquad \frac{mg}{mmol} \text{ (for conversions)}$$

Content data in mixed phases

c(i) = molar concentration of substance i in mol/L or mmol/L
a(i) = activity of substance i in mol/L
β(i) = mass concentration of substance i in µg/L or mg/L or g/L
x(i) = substance amount fraction (mole fraction)

Formulaic relationships

Molar concentration

$$c(i) = \frac{n(i)}{V} \quad \frac{mol}{L} \qquad \text{or also} \qquad \frac{10^3\ mmol}{L}$$

Activity

$$a(i) = c(i) - f(i) \qquad mol/L \text{ or } mmol/L$$

f(i) = activity coefficient

Mass concentration

$$\beta(i) = c(i) - M(i) \quad \frac{g}{L} \qquad \text{or also :} \quad \frac{10^3\ mg}{L} \quad \frac{10^6\ \mu g}{L}$$

Substance amount share

$$x(i) = \frac{n(i)}{n} \quad \frac{mol}{mol}$$

n = total amount of substance in the mixed phase

11.2 Ionic strength and activity

(Basics for chapter 7.5)

If a chemical compound is present in water in the form of ions - and this is the case for almost all inorganic compounds - its molar concentration does not correspond to its **effective** concentration in chemical reactions. The ions influence each other by their charges, so that the number of particles participating in the equilibrium reactions, and thus the *effective* concentration, is smaller than the molar concentration c(i).

The more ions there are in a water, the greater the mutual influence of the ions on each other. Because of their charge density, divalent or trivalent ions have a greater influence on the other ions than monovalent ions.

A measure of a water's ionic content is the ionic strength I.

Ionic strength

The equations for calculating the ionic strength are presented in chapter 7.5 .

$$I = 0.5 \cdot \sum_{j=1}^{j} [c(i) \cdot (z(i)^2)] \qquad [mol/L]$$

c(i) = concentration of the ion (i) [mol/L]
z(i) = charge number
j = number of ions

Example calculation of ionic strength with concentrations for a $0.2 \cdot 10^{-3}$ molar $CaCl_2$ solution :

$c(Ca^{2+}) = 0.2 \cdot 10^{-3}$ mol/L $\qquad c(Cl^-) = 0.4 \cdot 10^{-3}$ mol/L

$z(Ca^{2+}) = 2 \qquad z(Cl^-) = 1$

$I = 0.5 \cdot [(0.2 \cdot 10^{-3} \cdot 2^2) + (0.4 \cdot 10^{-3} \cdot 1^2)]$

$I = 0.6 \cdot 10^{-3}$ mol/L

Activity and activity coefficient

When analyzing substances in water, one always determines the total content of the substance → the molar concentration. For the evaluation of equilibrium reactions, on the other hand, only the active amount of the substance is of importance.

The *effective* concentration is called the activity a(i) of substance i.

$$a(i) < c(i)$$

Activity and concentration of a specific ion (i) are linked as follows:

$$a(i) = c(i) \cdot f(i) \qquad\qquad f(i) = \text{activity coefficient}$$

The activity coefficient f(i) can take values between $0 < f \leq 1$.

Only in ideally diluted solutions it can be 1, in concentrated solutions the activity coefficient is smaller than 1. Thus, as an approximation, only in highly dilute solutions can the activity of a substance be set approximately equal to its concentration:

Approximation only in highly dilute solutions:

$$a(i) \approx c(i)$$

In concentrated solutions, the activity coefficient can be much smaller than 1, i.e., the activity of the ions can be severely limited.

The activity coefficient f(i) depends on the intensity of mutual influence of ions among themselves, which is expressed by the ionic strength I.

For the dependence of the activity coefficient of a specific ion (i) on the ionic strength I, the laws according to DEBEYE-HÜCKEL apply:

$$\lg(f(i)) = -A \cdot (z(i))^2 \cdot \frac{\sqrt{I}}{1+B \cdot a(i) \cdot \sqrt{I}} \qquad\qquad I \text{ in mol/L !}$$

A and B are characteristic, temperature-dependent constants for the solvent (see also chapter 7.5).

11.3 The law of mass action

Using the dissociation reaction of an acid AH as an example :

$$AH \underset{v_2}{\overset{v_1}{\rightleftharpoons}} A^- + H^+$$

(A^- = acid anion resp. "acid residue")
(H^+ simplified = H_3O^+)
(reaction velocities v_1 and v_2)

This reaction is a chemical equilibrium reaction. In the (dynamic !) **Equilibrium**, the reaction rates of the outward reaction (v_1) and the reverse reaction (v_2) are equal and their effects cancel each other out. The molar concentrations of the reactants remain unchanged. The reaction rate in each case is directly dependent on the activities of the reaction partners:

$$v_1 = k_1 \cdot a(AH)$$
$$v_2 = k_2 \cdot a(A^-) \cdot a(H^+)$$

where k_1 and k_2 are the rate constants of the reaction. At $v_1 = v_2$, i.e. at equilibrium, the following applies:

$$v_1 = v_2 = k_1 \cdot a(AH) = k_2 \cdot a(A^-) \cdot a(H^+)$$

If one forms a new constant K from the two velocity constants, it follows:

$$\frac{k_1}{k_2} = K = \frac{a(A^-) \cdot a(H^+)}{a(AH)}$$

or with words: The mathematical product of the **activities** of the reaction products divided by the product of the **activities** of the starting substances is constant (law of mass action).

$$\frac{a(A^-) \cdot a(H^+)}{a(AH)} = K$$

K = mass action constant
or equilibrium constant
or dissociation constant of the acid

The mass action law in this form is independent of the ionic strength of the water, since it is derived directly from the activities of the substances. The mass action constant K depends on the pressure and temperature.

<u>Consideration of the ionic strength</u>

Similarly, a mass action constant K* can be defined for the molar concentrations of the substances:

$$\frac{c(A^-) \cdot c(H^+)}{c(AH)} = K^*$$

This mass action constant K* depends on the ionic strength I in addition to pressure and temperature. It applies:

$$K = K^* \cdot f_K$$

Derivation of the correlation:

The reaction rates v_1 and v_2 depend on the activities a(i) of the substances involved (i). Only in the hypothetical case of an extremely dilute solution with an ionic strength approaching zero can be assumed:

$$a(i) \approx c(i) \qquad \text{(I approximated = 0)}$$

An increasing ionic strength leads to a decrease in the activity of a substance (i) if the concentration of this substance remains unchanged. This is taken into account by the activity coefficient f.

$$a(i) = c(i) \cdot f(i)$$

The relationship between the constants K and K* is obtained by inserting the product of concentration and activity coefficient for the activities of the substances.

$$\frac{a(A^-) \cdot a(H^+)}{a(AH)} = \frac{c(A^-) \cdot f(A^-) \cdot c(H^+) \cdot f(H^+)}{c(AH) \cdot f(AH)} = K$$

$$\frac{c(A^-) \cdot f(A^-) \cdot c(H^+) \cdot f(H^+)}{c(AH) \cdot f(AH)} = \frac{c(A^-) \cdot c(H^+)}{c(AH)} \cdot \frac{f(A^-) \cdot f(H^+)}{f(AH)} = K^* \frac{f(A^-) \cdot f(H^+)}{f(AH)} = K$$

$$\rightarrow \quad K = K^* \cdot f_K \qquad \text{with } f_K = \frac{f(A^-) \cdot f(H^+)}{f(AH)}$$

If H_2O is present in a reaction equation, the activity $a(H_2O) = c(H_2O)$ (f=1 !) is considered constant and is usually included in the mass action constant.

Example: Dissociation constant K_1 of carbonic acid (chap. 4.5)

$$CO_2 + H_2O \rightleftharpoons HCO_3^- + H^+$$

$$\frac{a(H^+) \cdot a(HCO_3^-)}{a(CO_2) \cdot a(H_2O)} = K$$

$$\frac{a(H^+) \cdot a(HCO_3^-)}{a(CO_2)} = K \cdot a(H_2O) = K_1$$

11.4 pH value

Because of the fundamental importance of pH in the carbonic acid system, the basics of pH will be presented in some detail.

Autoprotolysis of water (ion product)

Water dissociates (protolysis) to a small extent into H^+ and OH^- ions, which are not free in aqueous solution but are present in single or multiple hydrated form $(i)_{(aq)}$.

$$H_2O \rightleftharpoons H^+_{(aq)} + OH^-_{(aq)}$$

e.g. $H^+_{(aq)}$: H_3O^+, $H_5O_2^+$, $H_7O_3^+$, ...

This means that the reaction equation should actually read:

$$H_2O + H_2O \rightleftharpoons H_3O^+ + OH^-$$

However, for the sake of simplicity, one writes:

$$H_2O \rightleftharpoons H^+ + OH^-$$

The law of mass action is then for this reaction:

$$\frac{a(H^+) \cdot a(OH^-)}{a(H_2O)} = K$$

Since the small amount of dissociation (very small value for the constant K) means that the activity of the water can be considered constant, it is included in the constant.

$$c(H_2O) = 55{,}41 \text{ mol/L}$$

$$c(H^+) = 10^{-7} \text{ mol/L}$$

$$\rightarrow \quad a(H_2O) \approx \text{constant}$$

$$a(H^+) \cdot a(OH^-) = K \cdot a(H_2O) = K_w$$

The equation is then **(ion product of water)**:

$$a(H^+) \cdot a(OH^-) = K_w \qquad [(\text{mol/L})^2]$$

The product of $a(H^+)$ and $a(OH^-)$ is called the ion product of water. The value for the ion product depends on the temperature (chap. 7.1).

At **<u>25 °C</u>** applies : $K_w = 10^{-14}$ $(\text{mol/L})^2$

In **pure water** , 50% H^+ ions and 50% OH^- ions are always present due to dissociation.

$$H_2O \rightleftharpoons H^+ + OH^-$$
$$\mathbf{50\%} \quad \mathbf{50\%}$$

$$\rightarrow \quad a(H^+) = a(OH^-) \quad \rightarrow \quad [a(H^+)]^2 = K_w$$

$$a(H^+) = \sqrt{K_w}$$

$$a(OH^-) = \sqrt{K_w}$$

$$\sqrt{K_w} = \sqrt{10^{-14}} = 10^{-7} \qquad (25\ °C)$$

There are then 10^{-7} mol/L H^+ ions
and 10^{-7} mol/L OH^- ions are present in the water **(25 °C** !).

Such water with equal concentrations of H^+ and OH^- ions is called ***neutral*** .

If more H^+ ions are present than OH^- ions, it is classified as ***acidic***, if more OH^- ions are present, it is classified as ***alkaline (basic)*** .

pH value: Definition pH/pOH

In order to characterize the water properties described above and to obtain simpler values, pH value has been defined as follows :

$$pH = - lg\ a(H^+)$$

! The pH value is the negative decadic logarithm
of the hydrogen ion **activity** $a(H^+)$ **!**

$a(H^+)$ in **mol/L !!!**

During pH measurement, only the active portion is recorded!

The operator p is a calculation rule which means: form the negative decadic logarithm. The operator p is used more frequently in chemistry.

$$p(i) \equiv - lg(i)$$

so : **(at 25 °C !)**
$a(H^+) = 10^{-7}$ mol/l → pH = 7 = neutral
$a(H^+) > 10^{-7}$ mol/l → pH < 7 = acidic
$a(H^+) < 10^{-7}$ mol/l → pH > 7 = basic or alkaline

To simplify calculations, one has defined analogous to pH using the same operator :

$$pK_w = -\lg K_w$$

and $$pOH = -\lg a(OH^-)$$

This results in : $$pK_w = pH + pOH$$

$$14 = 7 + 7 \qquad \text{(at 25 °C !)}$$

Temperature dependence of the pH value

A measured pH value is only valid for the temperature at which the measurement was performed! With a change in temperature, the ion product of the water and thus $a(H^+)$ also changes.

Meaning : Specification of a pH measured value always with temperature specification!

The **neutral point** also depends on the temperature. It is defined like this,

that $a(H^+) = a(OH^-)$ or $pH = pOH$

and thus $pH = pK_w/2$

Table 11.4-1: Temperature dependence of the pH value

Temp. °C	pKw	pH neutral
0	14.944	7.47
5	14.834	7.37
10	14.535	7.27
15	14.346	7.17
20	14.166	7.08
25	13.996	7.00

Only at 25 °C neutral water has a pH value of 7.0 !

! As the temperature rises, the pH value falls !

Annotation:

The term 'neutral' is only related to the state pH = pOH and does not contain any statement about the behavior of this water towards materials!

A neutral drinking water (pH = 7) is usually lime dissolving!

pH indicators

pH indicators are only discussed here because they were historically the basis for determining acid and base capacity and gave their names to m-value (methyl orange) and p-value (phenolphthalein).

pH indicators are organic dyes that change color (weak acids or bases) by absorbing or releasing protons (H^+ ions). For pH indicators, a changeover point for the color change is specified.

Indicator	Turnover range (pH)	Color change
Methyl orange	3.1 - 4.4	red - yellow orange
Phenolphthalein	8.2 - 9.8	colorless - red

Measurement of the pH value

The pH value is measured with pH electrodes. It should be noted here that special soft water electrodes must be used for very soft water or rainwater.

Although the pH measurement appears to be primarily simple, fundamental errors are quite often made here and incorrect values are determined. However, since the pH value is a basic parameter in the carbonic acid system, special care must be taken here.

Calibration :

The pH electrode must be calibrated before use.

The pH values of the calibration buffers are temperature-dependent and are only valid for the specified temperature, usually 25 °C !

The operating manuals for the various pH measuring instruments contain special calibration instructions.

Principle procedure

Sampling in keywords

- as little air contact as possible - especially with soft water
- measure as directly as possible (temperature ! , possibly in the flow)
- otherwise transport cool (outgassing !)

Carrying out the measurement

rinse pH electrode (and temperature sensor)

best with sample water, then no dabbing is necessary

otherwise rinse with deion. water and dab lightly

immerse the pH electrode (and temperature sensor) in the test sample and wait for stable pH display

Indication of the measurement results:

For absolute measurements at least one digit after the decimal point, the measuring temperature must also be specified.
For relative measurements, the specification of two decimal places is also useful.

Example: pH value 7.46 t_{pH} = 12.3 °C

11.5 Acids

It is the characteristic of acids that they dissociate in aqueous solution and release H^+ ions to the water (definition according to BRÖNSTED):

$$AH \rightleftharpoons A^- + H^+$$

(A^-= acid anion resp. "acid residue")
(H^+ simplified = H_3O^+)

The mass action equation derived from this reaction is:

$$\frac{a(A^-)\cdot a(H^+)}{a(AH)} = K$$

K = mass action constant
or equilibrium constant
or dissociation constant of the acid

In analogy to the definition of the pH value, the negative decadic logarithm of the mass action constant K is defined as the pK value. To indicate that these are dissociation constants of acids, the subscript S (B for bases) is added:

$$pK = -\lg(K) \quad \longrightarrow \quad pK_S = -\lg(K_S)$$

The acid constant K_S / base constant K_B must not be confused with the abbreviation $K_{S4.3}$ for the acid capacity / base capacity $K_{B8.2}$ (see 7.2.2)!

The strength of an acid is the extent of dissociation. Nearly completely dissociated acids have high acid constants $K > 1$ and are called "strong":

strong acids	$K > 1$	$pK < 0$
medium strong acids	$1 > K > 10^{-4}$	$0 < pK < 4$
weak acids	$K < 10^{-4}$	$4 < pK$

Strong mineral acids generally exist completely dissociated in aqueous solutions.

<u>Carbonic acid as a weak acid (simplified carbonic acid system) :</u>

Temperature in °C	K_1 in mol/L	K_2 in mol/L
10	$3.426 \cdot 10^{-7}$	$3.249 \cdot 10^{-11}$
25	$4.406 \cdot 10^{-7}$	$4.688 \cdot 10^{-11}$

The following pKs values are obtained from the dissociation constants :

Acid name	Reaction	pK_S value	
		10 °C	25 °C
Carbon dioxide	$CO_2 + H_2O \rightleftharpoons H^+ + HCO_3^-$	6.47	6.36
Hydrogen carbonation	$HCO_3^- \rightleftharpoons H^+ + CO_3^{2-}$	10.49	10.33

If the pK_S value is equal to the pH of the water, the acid is 50% dissociated.

Table 11.5-1: Classification of acids according to acid strength

Division	Acid name	Reaction	pKs value
strong acids	Hydrochloric acid	$HCl \leftrightharpoons H^+ + Cl^-$	-6
	Sulfuric acid	$H_2SO_4 \leftrightharpoons H^+ + HSO_4^-$	-3
	Nitric acid	$HNO_3 \leftrightharpoons H^+ + NO_3^-$	-1.34
weak acids	Acetic acid	$CH_3COOH \leftrightharpoons H^+ + CH_3COO^-$	4.75
	Carbon dioxide	$CO_2 + H_2O \leftrightharpoons H^+ + HCO_3^-$	6.36
	Hydrogen carbonation	$HCO_3^- \leftrightharpoons H^+ + CO_3^{2-}$	10.33
	Phosphoric acid	$H_3PO_4 \leftrightharpoons H^+ + H_2PO_4^-$	2.12
	Dihydrogen phosphate	$H_2PO_4^- \leftrightharpoons H^+ + HPO_4^{2-}$	7.21
	Hydrogen phosphate	$HPO_4^{2-} \leftrightharpoons H^+ + PO_4^{3-}$	12.67
	Ammonium ion	$NH_4^+ \leftrightharpoons H^+ + NH_3$	9.25
	Silica	$H_4SiO_4 \leftrightharpoons H^+ + H_3SiO_4^-$	9.6

11.6 Buffering

Buffering is the ability of an aqueous solution to respond to the addition of even large amounts of strong acid or base with only a slight change in pH.

Most of the added H^+ ions (OH^- ions) are consumed by reaction with the buffering substances and do not cause any pH change. If one adds appropriate amounts of acid (or base) to pure water, the pH changes directly in full accordance with the amount of substance added.

A buffer solution **synthetically** produced in chemistry consists of a weak acid AH (base) and the salt of the weak acid (base) with the corresponding anion A^- (cation B^+), e.g. a phosphate buffer

$$H_2PO_4^- - HPO_4^{2-}$$

$$H_3PO_4 \underset{\nwarrow H^+}{\overset{OH^- \searrow}{\rightleftharpoons}} H_2PO_4^- + H^+ \qquad pK = 2{,}12 \quad (25\,°C)$$

$$H_2PO_4^- \underset{\nwarrow H^+}{\overset{OH^- \searrow}{\rightleftharpoons}} HPO_4^{2-} + H^+ \qquad pK = 7{,}21 \quad (25\,°C)$$

Addition of H^+ leads to the formation of the undissociated acid.
The addition of OH- leads to the consumption of H^+ via the neutralization reaction ($OH^- + H^+ \rightarrow H_2O$) and thus to the dissociation of the acid.

Table 11.6-1: Natural buffer substances

Buffer - pair	pKs value		Source
CO_2 - HCO_3^-	6.36	(25 °C)	[44]
$H_2PO_4^-$ - HPO_4^{2-}	7.21	(25 °C)	[82]
H_3BO_3 (Seawater)	8.85	(10 °C)	[44]
NH_4^+ - NH_3	9.3	(25 °C)	[82]
H_4SiO_4 - $H_3SiO_4^-$	9.6	(25 °C)	[82]
HCO_3^- - CO_3^{2-}	10.33	(25 °C)	[44]

In **natural waters**, buffering properties are caused by weak acids or weak bases, which are partly dissociated (pKs values in the range 4 - 11). As a rule, carbonic acid has a buffering effect in natural waters. High humic acid contents can have an additional buffering effect. In phosphated drinking waters, the additional buffering effect of phosphate must be taken into account. Seawater contains

$4.1 \cdot 10^{-4}$ mol/L of boric acid and borates, but these have little effect on the buffering intensity caused predominantly by carbonic acid [44].

The buffering intensity is greatest when pH = pK_S value, since then $a(A^-) = a(AH)$.

It applies: $$pH = pK_S + \lg \frac{a(A^-)}{a(AH)}$$ (so-called buffer equation) (HENDERSON-HASSELBALCH)

<u>Derivation of the equation:</u>

According to the definition is:

$pK_S = -\lg K_S$ (see acid strength) and $pH = -\lg a(H^+)$

Transformation of the mass action equation :

$$K_S = \frac{a(A^-) \cdot a(H^+)}{a(AH)}$$

$$\lg K_S = \lg \frac{a(A^-)}{a(AH)} + \lg a(H^+)$$

$$-\lg a(H^+) = -\lg K_S + \lg \frac{a(A^-)}{a(AH)}$$

$$pH = pK_S + \lg \frac{a(A^-)}{a(AH)}$$

$$\Rightarrow \quad pH = pK_S \quad \text{if} \quad \lg \frac{a(A^-)}{a(AH)} = 0 \quad \Rightarrow \quad a(A^-) = a(AH)$$

Thus, when the dissociated (A^-) and undissociated (AH) portions of the weak acid are equal, the buffering effect is greatest. This is the case when the pH value corresponds to the pK value of the acid :

$$pH = pK_S$$

11.7 The solubility product

Due to ionic bonding, salts are also present in the crystal lattice as ions. The dissolution process of salts in water is therefore a purely physical process.

The solubility of salts is decisively determined by the size of their lattice energy and hydration energy of their ions. Most salts have a high solubility in water, which usually increases with increasing temperature. However, there are exceptions, e.g. calcite: the solubility decreases with increasing temperature.

However, some salts dissolve in water only in low concentrations (sparingly soluble).

Reaction equation: $\downarrow AB_{(solid)} \rightleftharpoons A^- + B^+$

The law of mass action applies to this equation.

$$\frac{a(A^-) \cdot a(B^+)}{a(AB)_{(solid)}} = K$$

In the reaction equilibrium, the outward and reverse reactions - solution and precipitation - proceed at the same rate; a saturated solution is present.

$$\downarrow AB_{(solid)} \underset{v_2}{\overset{v_1}{\rightleftharpoons}} A^- + B^+$$

$$v_1 = v_2 \quad \longrightarrow \quad \text{saturated solution}$$

Since the amount of solid AB has no influence on the equilibrium concentrations arising in the solution equilibrium, the activity of the soil body is set a(AB) = 1 and the associated constant is called the solubility product L:

$$a(A^-) \cdot a(B^+) = L \qquad [mol/L]^2$$

The product of the ion activities must reach the value of the solubility product so that the maximum solubility of AB is reached. If it is exceeded, this compound precipitates out of solution.

$a(A^-) \cdot a(B^+) < L$ = soluble
$a(A^-) \cdot a(B^+) = L$ = solubility limit
$a(A^-) \cdot a(B^+) > L$ = solubility exceeded → Substance precipitates

In carbonic acid chemistry, the formula symbol K_C is commonly used for the solubility product of calcite:

$$K_C = L(\text{Calcit}) = a(Ca^{2+}) \cdot a(CO_3^{2-}) \qquad (mol/L)^2$$

12 Tables

12.1 Values for the HENRY constant

Calculation see chap. 7.1.1

Table 12.1-1: Values for the HENRY constant

	Freshwater I = 0	**Salt water** S=35.0‰ Cl=19.37‰
Temperature °C	K_H mol/(m³·bar)	K_H mol/(m³·bar)
0	76.895	62.303
1	73.897	60.029
2	71.060	57.868
3	68.374	55.815
4	65.830	53.863
5	63.418	52.006
6	61.131	50.239
7	58.960	48.556
8	56.898	46.953
9	54.940	45.425
10	53.077	43.969
11	51.306	42.580
12	49.620	41.254
13	48.014	39.989
14	46.484	38.781
15	45.026	37.626
16	43.635	36.523
17	42.308	35.468
18	41.041	34.460
19	39.830	33.494
20	38.674	32.571
21	37.568	31.686
22	36.510	30.839
23	35.498	30.027
24	34.529	29.249
25	33.600	28.503

12.2 Base constants as a function of temperature

Calculation basis see chap. 7.1

Table 12.2-1: Base constants as a function of temperature

Temperature	K_1	K_2	K_W	$L = K_C$
°C	mol/L	mol/L	mol/L	(mol/L)^2
0	$2.677 \cdot 10^{-7}$	$2.352 \cdot 10^{-11}$	$1.140 \cdot 10^{-15}$	$4.244 \cdot 10^{-9}$
1	$2.753 \cdot 10^{-7}$	$2.437 \cdot 10^{-11}$	$1.258 \cdot 10^{-15}$	$4.218 \cdot 10^{-9}$
2	$2.828 \cdot 10^{-7}$	$2.523 \cdot 10^{-11}$	$1.387 \cdot 10^{-15}$	$4.190 \cdot 10^{-9}$
3	$2.904 \cdot 10^{-7}$	$2.610 \cdot 10^{-11}$	$1.528 \cdot 10^{-15}$	$4.161 \cdot 10^{-9}$
4	$2.979 \cdot 10^{-7}$	$2.699 \cdot 10^{-11}$	$1.681 \cdot 10^{-15}$	$4.131 \cdot 10^{-9}$
5	$3.055 \cdot 10^{-7}$	$2.788 \cdot 10^{-11}$	$1.847 \cdot 10^{-15}$	$4.100 \cdot 10^{-9}$
6	$3.130 \cdot 10^{-7}$	$2.879 \cdot 10^{-11}$	$2.029 \cdot 10^{-15}$	$4.067 \cdot 10^{-9}$
7	$3.205 \cdot 10^{-7}$	$2.970 \cdot 10^{-11}$	$2.225 \cdot 10^{-15}$	$4.034 \cdot 10^{-9}$
8	$3.279 \cdot 10^{-7}$	$3.062 \cdot 10^{-11}$	$2.439 \cdot 10^{-15}$	$3.999 \cdot 10^{-9}$
9	$3.353 \cdot 10^{-7}$	$3.155 \cdot 10^{-11}$	$2.670 \cdot 10^{-15}$	$3.964 \cdot 10^{-9}$
10	$3.426 \cdot 10^{-7}$	$3.249 \cdot 10^{-11}$	$2.920 \cdot 10^{-15}$	$3.927 \cdot 10^{-9}$
11	$3.499 \cdot 10^{-7}$	$3.343 \cdot 10^{-11}$	$3.192 \cdot 10^{-15}$	$3.890 \cdot 10^{-9}$
12	$3.571 \cdot 10^{-7}$	$3.438 \cdot 10^{-11}$	$3.485 \cdot 10^{-15}$	$3.852 \cdot 10^{-9}$
13	$3.642 \cdot 10^{-7}$	$3.534 \cdot 10^{-11}$	$3.801 \cdot 10^{-15}$	$3.813 \cdot 10^{-9}$
14	$3.712 \cdot 10^{-7}$	$3.630 \cdot 10^{-11}$	$4.142 \cdot 10^{-15}$	$3.774 \cdot 10^{-9}$
15	$3.781 \cdot 10^{-7}$	$3.726 \cdot 10^{-11}$	$4.511 \cdot 10^{-15}$	$3.733 \cdot 10^{-9}$
16	$3.849 \cdot 10^{-7}$	$3.822 \cdot 10^{-11}$	$4.907 \cdot 10^{-15}$	$3.693 \cdot 10^{-9}$
17	$3.916 \cdot 10^{-7}$	$3.919 \cdot 10^{-11}$	$5.335 \cdot 10^{-15}$	$3.651 \cdot 10^{-9}$
18	$3.982 \cdot 10^{-7}$	$4.015 \cdot 10^{-11}$	$5.794 \cdot 10^{-15}$	$3.609 \cdot 10^{-9}$
19	$4.047 \cdot 10^{-9}$	$4.112 \cdot 10^{-11}$	$6.288 \cdot 10^{-15}$	$3.567 \cdot 10^{-9}$
20	$4.110 \cdot 10^{-7}$	$4.209 \cdot 10^{-11}$	$6.817 \cdot 10^{-15}$	$3.524 \cdot 10^{-9}$
21	$4.172 \cdot 10^{-7}$	$4.305 \cdot 10^{-11}$	$7.386 \cdot 10^{-15}$	$3.480 \cdot 10^{-9}$
22	$4.233 \cdot 10^{-7}$	$4.402 \cdot 10^{-11}$	$7.995 \cdot 10^{-15}$	$3.437 \cdot 10^{-9}$
23	$4.292 \cdot 10^{-7}$	$4.498 \cdot 10^{-11}$	$8.648 \cdot 10^{-15}$	$3.393 \cdot 10^{-9}$
24	$4.350 \cdot 10^{-7}$	$4.593 \cdot 10^{-11}$	$9.346 \cdot 10^{-15}$	$3.348 \cdot 10^{-9}$
25	$4.406 \cdot 10^{-7}$	$4.688 \cdot 10^{-11}$	$1.010 \cdot 10^{-14}$	$3.304 \cdot 10^{-9}$

12.3 Proportions of carbonic acid forms as a function of pH value

Table 12.3-1: Proportions of carbonic acid forms as a function of pH (I=0)

	10 °C				25 °C		
pH	$x(CO_2)$	$x(HCO_3^-)$	$x(CO_3^{2-})$	pH	$x(CO_2)$	$x(HCO_3^-)$	$x(CO_3^{2-})$
5.0	0.967	0.0331	$0.108 \cdot 10^{-6}$	5.0	0.958	0.0422	$0.198 \cdot 10^{-6}$
5.2	0.948	0.0515	$0.265 \cdot 10^{-6}$	5.2	0.935	0.0653	$0.485 \cdot 10^{-6}$
5.4	0.921	0.0792	$0.647 \cdot 10^{-6}$	5.4	0.900	0.0996	$0.117 \cdot 10^{-5}$
5.6	0.880	0.120	$0.155 \cdot 10^{-5}$	5.6	0.851	0.149	$0.278 \cdot 10^{-5}$
5.8	0.822	0.178	$0.364 \cdot 10^{-5}$	5.8	0.782	0.218	$0.643 \cdot 10^{-5}$
6.0	0.745	0.255	$0.829 \cdot 10^{-5}$	6.0	0.694	0.306	$0.143 \cdot 10^{-4}$
6.2	0.648	0.352	$0.181 \cdot 10^{-4}$	6.2	0.589	0.411	$0.305 \cdot 10^{-4}$
6.4	0.537	0.463	$0.378 \cdot 10^{-4}$	6.4	0.475	0.525	$0.619 \cdot 10^{-4}$
6.6	0.423	0.577	$0.746 \cdot 10^{-4}$	6.6	0.363	0.637	$0.119 \cdot 10^{-3}$
6.8	0.316	0.684	$0.140 \cdot 10^{-3}$	6.8	0.265	0.735	$0.217 \cdot 10^{-3}$
7.0	0.226	0.774	$0.251 \cdot 10^{-3}$	7.0	0.185	0.815	$0.382 \cdot 10^{-3}$
7.2	0.155	0.844	$0.435 \cdot 10^{-3}$	7.2	0.125	0.874	$0.650 \cdot 10^{-3}$
7.4	0.104	0.895	$0.731 \cdot 10^{-3}$	7.4	$0.828 \cdot 10^{-3}$	0.916	$0.108 \cdot 10^{-2}$
7.6	$0.682 \cdot 10^{-1}$	0.931	$0.120 \cdot 10^{-2}$	7.6	$0.538 \cdot 10^{-1}$	0.944	$0.176 \cdot 10^{-2}$
7.8	$0.441 \cdot 10^{-1}$	0.954	$0.196 \cdot 10^{-2}$	7.8	$0.346 \cdot 10^{-1}$	0.963	$0.285 \cdot 10^{-2}$
8.0	$0.283 \cdot 10^{-1}$	0.969	$0.315 \cdot 10^{-2}$	8.0	$0.221 \cdot 10^{-1}$	0.973	$0.456 \cdot 10^{-2}$
8.2	$0.180 \cdot 10^{-1}$	0.977	$0.503 \cdot 10^{-2}$	8.2	$0.140 \cdot 10^{-1}$	0.979	$0.727 \cdot 10^{-2}$
8.4	$0.114 \cdot 10^{-1}$	0.981	$0.800 \cdot 10^{-2}$	8.4	$0.885 \cdot 10^{-2}$	0.980	$0.115 \cdot 10^{-1}$
8.6	$0.719 \cdot 10^{-2}$	0.980	$0.127 \cdot 10^{-1}$	8.6	$0.557 \cdot 10^{-2}$	0.976	$0.182 \cdot 10^{-1}$
8.8	$0.451 \cdot 10^{-2}$	0.975	$0.200 \cdot 10^{-1}$	8.8	$0.348 \cdot 10^{-2}$	0.968	$0.286 \cdot 10^{-1}$
9.0	$0.282 \cdot 10^{-2}$	0.966	$0.314 \cdot 10^{-1}$	9.0	$0.216 \cdot 10^{-2}$	0.953	$0.447 \cdot 10^{-1}$

12.4 Constants of the complexation of calcium

$Ca^{2+} + HCO_3^-$	⇌	$CaHCO_3^+$	K_3
$Ca^{2+} + CO_3^{2-}$	⇌	$CaCO_3^0$	K_4
$Ca^{2+} + OH^-$	⇌	$Ca(OH)^+$	K_5
$Ca^{2+} + H_2PO_4^-$	⇌	$CaH_2PO_4^+$	K_6
$Ca^{2+} + HPO_4^{2-}$	⇌	$CaHPO_4$	K_7
$Ca^{2+} + PO_4^{3-}$	⇌	$CaPO_4^-$	K_8
$Ca^{2+} + SO_4^{2-}$	⇌	$CaSO_4^0$	K_9

°C	K_3	K_4	K_5	K_6	K_7	K_8	K_9
0	9.852E+00	1.077E+03	1.386E+01	4.365E+00	1.855E+02	1.895E+06	1.672E+02
1	9.978E+00	1.086E+03	1.413E+01	4.593E+00	1.931E+02	1.930E+06	1.681E+02
2	1.011E+01	1.097E+03	1.440E+01	4.825E+00	2.007E+02	1.966E+06	1.691E+02
3	1.023E+01	1.108E+03	1.467E+01	5.060E+00	2.085E+02	2.002E+06	1.701E+02
4	1.036E+01	1.119E+03	1.495E+01	5.300E+00	2.163E+02	2.039E+06	1.712E+02
5	1.049E+01	1.132E+03	1.523E+01	5.542E+00	2.241E+02	2.076E+06	1.723E+02
6	1.062E+01	1.145E+03	1.552E+01	5.786E+00	2.320E+02	2.113E+06	1.734E+02
7	1.075E+01	1.160E+03	1.581E+01	6.033E+00	2.399E+02	2.151E+06	1.746E+02
8	1.088E+01	1.175E+03	1.610E+01	6.280E+00	2.478E+02	2.189E+06	1.758E+02
9	1.102E+01	1.191E+03	1.640E+01	6.529E+00	2.558E+02	2.227E+06	1.771E+02
10	1.115E+01	1.208E+03	1.671E+01	6.777E+00	2.637E+02	2.266E+06	1.784E+02
11	1.128E+01	1.225E+03	1.702E+01	7.026E+00	2.716E+02	2.305E+06	1.798E+02
12	1.142E+01	1.244E+03	1.733E+01	7.273E+00	2.795E+02	2.345E+06	1.812E+02
13	1.155E+01	1.264E+03	1.765E+01	7.519E+00	2.873E+02	2.385E+06	1.827E+02
14	1.169E+01	1.284E+03	1.797E+01	7.763E+00	2.951E+02	2.426E+06	1.842E+02
15	1.182E+01	1.306E+03	1.830E+01	8.004E+00	3.028E+02	2.466E+06	1.858E+02
16	1.196E+01	1.329E+03	1.864E+01	8.241E+00	3.104E+02	2.508E+06	1.874E+02
17	1.209E+01	1.352E+03	1.897E+01	8.475E+00	3.179E+02	2.549E+06	1.891E+02
18	1.223E+01	1.377E+03	1.932E+01	8.704E+00	3.253E+02	2.592E+06	1.908E+02
19	1.237E+01	1.403E+03	1.967E+01	8.929E+00	3.326E+02	2.634E+06	1.925E+02
20	1.251E+01	1.430E+03	2.002E+01	9.147E+00	3.398E+02	2.677E+06	1.943E+02
21	1.265E+01	1.459E+03	2.038E+01	9.360E+00	3.468E+02	2.720E+06	1.962E+02
22	1.279E+01	1.488E+03	2.075E+01	9.566E+00	3.537E+02	2.764E+06	1.981E+02
23	1.293E+01	1.519E+03	2.112E+01	9.765E+00	3.604E+02	2.808E+06	2.001E+02
24	1.307E+01	1.551E+03	2.150E+01	9.956E+00	3.669E+02	2.852E+06	2.021E+02
25	1.321E+01	1.585E+03	2.188E+01	1.014E+01	3.733E+02	2.897E+06	2.042E+02

12.5 Constants of the complexation of magnesium

$Mg^{2+} + HCO_3^-$	⇌	$MgHCO_3^+$	K_{10}
$Mg^{2+} + CO_3^{2-}$	⇌	$MgCO_3^0$	K_{11}
$Mg^{2+} + OH^-$	⇌	$Mg(OH)^+$	K_{12}
$Mg^{2+} + H_2PO_4^-$	⇌	$MgH_2PO_4^+$	K_{13}
$Mg^{2+} + HPO_4^{2-}$	⇌	$MgHPO_4$	K_{14}
$Mg^{2+} + PO_4^{3-}$	⇌	$MgPO_4^-$	K_{15}
$Mg^{2+} + SO_4^{2-}$	⇌	$MgSO_4^0$	K_{16}

°C	K_{10}	K_{11}	K_{12}	K_{13}	K_{14}	K_{15}	K_{16}
0	8.948E+00	5.662E+02	1.344E+02	1.380E+01	3.551E+02	2.402E+06	9.472E+01
1	9.053E+00	5.766E+02	1.354E+02	1.454E+01	3.705E+02	2.452E+06	9.746E+01
2	9.158E+00	5.871E+02	1.364E+02	1.529E+01	3.861E+02	2.504E+06	1.003E+02
3	9.264E+00	5.978E+02	1.374E+02	1.605E+01	4.019E+02	2.556E+06	1.031E+02
4	9.370E+00	6.087E+02	1.385E+02	1.683E+01	4.179E+02	2.609E+06	1.060E+02
5	9.477E+00	6.198E+02	1.395E+02	1.762E+01	4.340E+02	2.662E+06	1.090E+02
6	9.584E+00	6.311E+02	1.406E+02	1.842E+01	4.504E+02	2.716E+06	1.121E+02
7	9.691E+00	6.425E+02	1.418E+02	1.923E+01	4.668E+02	2.771E+06	1.152E+02
8	9.799E+00	6.542E+02	1.429E+02	2.005E+01	4.833E+02	2.827E+06	1.184E+02
9	9.908E+00	6.661E+02	1.441E+02	2.087E+01	4.999E+02	2.883E+06	1.216E+02
10	1.002E+01	6.781E+02	1.452E+02	2.169E+01	5.166E+02	2.940E+06	1.249E+02
11	1.013E+01	6.904E+02	1.465E+02	2.252E+01	5.333E+02	2.998E+06	1.283E+02
12	1.024E+01	7.028E+02	1.477E+02	2.334E+01	5.500E+02	3.056E+06	1.318E+02
13	1.035E+01	7.155E+02	1.489E+02	2.417E+01	5.666E+02	3.115E+06	1.353E+02
14	1.046E+01	7.284E+02	1.502E+02	2.499E+01	5.832E+02	3.175E+06	1.389E+02
15	1.057E+01	7.415E+02	1.515E+02	2.580E+01	5.998E+02	3.236E+06	1.425E+02
16	1.068E+01	7.548E+02	1.529E+02	2.660E+01	6.162E+02	3.297E+06	1.462E+02
17	1.079E+01	7.684E+02	1.542E+02	2.740E+01	6.325E+02	3.359E+06	1.501E+02
18	1.090E+01	7.821E+02	1.556E+02	2.818E+01	6.486E+02	3.422E+06	1.539E+02
19	1.101E+01	7.961E+02	1.570E+02	2.895E+01	6.646E+02	3.485E+06	1.579E+02
20	1.113E+01	8.104E+02	1.584E+02	2.971E+01	6.803E+02	3.549E+06	1.619E+02
21	1.124E+01	8.248E+02	1.599E+02	3.044E+01	6.958E+02	3.614E+06	1.660E+02
22	1.135E+01	8.395E+02	1.614E+02	3.116E+01	7.111E+02	3.680E+06	1.702E+02
23	1.147E+01	8.545E+02	1.629E+02	3.186E+01	7.261E+02	3.746E+06	1.745E+02
24	1.158E+01	8.697E+02	1.644E+02	3.254E+01	7.407E+02	3.814E+06	1.788E+02
25	1.169E+01	8.851E+02	1.660E+02	3.319E+01	7.551E+02	3.882E+06	1.832E+02

13 Formula symbols and abbreviations

Substance content

n(i) = molar quantity of substance i in mol

a(i) = activity of substance i in mol/L

c(i) = molar concentration of substance i in mol/L or mmol/L

β(i) = mass concentration of substance i in µg/L or mg/L or g/L

M(i) = Molar mass of substance i in g/mol

x(i) = substance amount fraction (mole fraction)

c(ti) = total concentration of a substance including complexes

vol% = Volume fraction of a gas in %

ppmv = volume fraction parts per million

Indexes:

$(\,)_{(aq)}$ = aquatic

$(\,)_{(g)}$ = gaseous

$(\,)_{(w)}$ = dissolved in water

Simplification: H^+ instead of H_3O^+

Abbreviations and symbols:

D	Calcite dissolving capacity / Calcite separation capacity
DIC	Dissolved Inorganic Carbon
DOC	Dissolved Organic Carbon
°dH	Degree of German hardness (1mmol/L Ca+Mg = 5.6 °dH)
Δ	$= c(OH^-) - c(H^+)$
ΔpH	$= pH - pH_C$
f	Activity coefficient for monovalent ions
f_2	Activity coefficient for divalent ions
f_3	Activity coefficient for trivalent ions
f_K	Activity coefficient to the constant K
F_{C1} F_{C2}	Complexing factors
F_{Ca}	Complexing factor Calcium
F_{Mg}	Complexing factor magnesium
F_{SO4}	Complexing factor sulfate
F_{HPO4}	Complexing factor Hydrogen phosphate
F_{H2PO4}	Complexing factor Dihydrogen phosphate
F_{PO4}	Complexing factor Phosphate
φ	Equivalence factor
I	Ionic strength (mol/L)
K_C	$L(CaCO_3)$

K_H	HENRY constant
K_W	Ion product of the water
K_1	1. Dissociation constant of carbonic acid
K_2	2. Dissociation constant of carbonic acid
K_3-K_{16}	Equilibrium constants in the carbonic acid system
$K_{S4.3}$	Acid capacity up to pH 4.3 (mol/L)
$K_{B8.2}$	Base capacity up to pH 8.2 (mol/L)
K*	Mass action constant with molar concentrations
κ_{25}	Electrical conductivity at 25 °C (mS/m)
L	Solubility product
LF	Electrical conductivity
m	m-value
p	p-value
p	Pressure
p_{ges}	Total pressure
pH	pH value
pH_A	Saturation pH value after carbon dioxide exchange
pH_C	calcite saturation pH
pH_L	Saturation pH value according to STROHECKER and LANGELIER
pH_t	pH of a water at measuring temperature t
pH_{tb}	pH of a water at evaluation temperature t_b
PI	Buffering intensity
SI	Saturation index
Sz^2	Sum of charge squares
T	Temperature in K
t	Temperature in °C
t_e	Withdrawal temperature in °C
t_b	Evaluation temperature in °C
t_{pH}	Temperature for pH measurement in °C
t_{tit}	Temperature during titration
TIC	Total Inorganic Carbon
TOC	Total Organic Carbon
vol%	Volume fraction of a gas in %
V_{mol}	Molar volume of a gas (L/mol)
$x(CO_2)$	Mass fraction for CO_2
$x(HCO_3^-)$	Mass fraction for HCO_3^-
$x(CO_3^{2-})$	Mass fraction for CO_3^{2-}
z(i)	Charge number

14 Bibliography

[1] Holleman, A.F. und Wiberg, E.: Lehrbuch der Anorganischen Chemie. W. de Gruyter Verlag, Berlin, 1976.

[2] „GESTIS-Stoffdatenbank, Institut für Arbeitsschutz der Deutschen Gesetzlichen Unfallversicherung (IFA), https://gestis.dguv.de/data?name=001120 (04.09.2021)".

[3] „https://webgate.ec.europa.eu/foods_system/main/index.cfm?event=substance.view&identifier=94&sector=FAD&auth=SANCAS (15.09.2021)".

[4] „https://www.cumschmidt.de/sm_co2.htm (07.09.2021)".

[5] Plummer, L.N. und Busenberg, E.: The solubilities of calcite, aragonite and vaterite in CO2-H2O solutions between 0 and 90 °C and an evaluation of the aqueous model for the system CaCO3-CO2-H2O. Geochimica et Cosmochimica Acta 46, S. 1011-1040, 1982.

[6] „Technische Regeln für Gefahrstoffe (TRGS 900): BArBl Heft 1/2006 S. 41-55 , zuletzt geändert und ergänzt: GMBl 2020, S. 902 [Nr. 42] (v. 27.10.2020)".

[7] „PubChem: https://pubchem.ncbi.nlm.nih.gov/compound/280#section=Health-Hazards (18.12.2020)".

[8] „Verordnung zum Schutz vor Gefahrstoffen (Gefahrstoffverordnung – GefStoffV), vom 26. November 2010 (BGBl. I S 1643), zuletzt geändert 29. März 2017 (BGBl. I S 626)".

[9] „DIN EN 378-1: Kälteanlagen und Wärmepumpen, Ausgabe Juni 2008 mit Berichtigung Januar 2010".

[10] Allred, A. L. and Rochow, E.G.: J. Inorg. Nucl. Chem., 1958, 5, 264.

[11] „Strahlungsphysik im optischen Bereich und Lichttechnik; Benennung der Wellenlängenbereiche. Deutsches Institut für Normung (Hrsg.): DIN. 5031 Teil 7, Januar 1984".

[12] „Lehmann, Th.: IR-Script, FU Berlin WS 2010, PDF, https://www.bcp.fu-berlin.de/chemie/chemie/studium/ocpraktikum/spektroskopie/unterlagen.html (22.12.2020)".

[13] „Czeslik, C., Seemann, H., and Winter, R., Basiswissen Physikalische Chemie, Vieweg und Teubner, GWV Fachverlage GmbH, Wiesbaden, 2010".

[14] „J. Weidlein, U. Müller, K. Dehnicke, Schwingungsspektroskopie, 2. Auflage, Thieme Stuttgart, 1988".

[15] „https://webbook.nist.gov/cgi/cbook.cgi?ID=C124389&Type=IR-SPEC&Index=1 (18.03.2022)".

[16] F. Alt, Schilfgras statt Atom, Piper-Verlag, 1992.

[17] „https://unfccc.int/process-and-meetings/the-kyoto-protocol/what-is-the-kyoto-protocol/kyoto-protocol-targets-for-the-first-commitment-period (03.05.2021)".

[18] „Umweltbundesamt:https://www.umweltbundesamt.de/themen/klima-energie/klimaschutz-energiepolitik-in-deutschland/treibhausgas-emissionen/die-treibhausgase#Methan (18.12.2020)".

[19] Myhre,G: et al.: Climate Change 2013: The Physical Science Basis. Working Group I contribution to the IPCC Fifth Assessment Report. Hrsg.: Intergovernmental Panel on Climate Change. 30. September 2013, Chapter 8: Anthropogenic and Natural Radiative Forcin.

[20] https://www.epa.gov/ghgemissions/understanding-global-warming-potentials (21.01.2023).

[21] „https://www.umweltbundesamt.de/daten/klima/treibhausgas-emissionen-in-deutschland#entwicklung-der-f-gase-teil-fluorierte-kohlenwasserstoffe-schwefelhexafluorid-und-stickstofftrifluorid (28.07.2022)".

[22] „Umweltbundesamt: https://uba.co2-rechner.de/de_DE/ (27.12.2020)".

[23] „Guido Reinhardt, Sven Gärtner, Tobias Wagner: Ökologische Fußabdrücke von Lebensmitteln und Gerichten in Deutschland. pdf, ifeu, Heidelberg, 2020. Link: https://www.google.de/url?sa=t&rct=j&q=&esrc=s&source=web&cd=&ved=2ahUKEwin4_zmupDzAhXj_rsIHUsWC2IQFno".

[24] „https://www.beuth.de/de/norm/din-en-iso-14067/289443505 (31.07.2022)".

[25] „Lee, C. A. et al.: A Framework for Understanding Whole-Earth Carbon Cycling. In: Orcutt. B. N. et al. ed.: Deep Carbon, Cambridge University Press, 2019".

[26] „Buchal, C. und Schönwiese, C-D.: Klima - Die Erde und ihre Atmosphäre im Wandel der Zeiten. MIC Agentur & Verlag; 3. Auflage 2016".

[27] „https://deepcarbon.net/".

[28] „Orcutt B, et al. eds. (2019) Deep Carbon: Past to Present. Cambridge, UK: Cambridge University Press doi: 10.1017/9781108677950".

[29] National Oceanic and Atmospheric Administration (NOAA) ftp://aftp.cmdl.noaa.gov/products/trends/co2/co2_annmean_mlo.txt.

[30] „https://de.statista.com/statistik/daten/studie/153731/umfrage/reserven-an-graphit-nach-laendern/ (25.09.2022)".

[31] National Oceanic and Atmospheric Administration (NOAA), ftp://aftp.cmdl.noaa.gov/products/trends/co2/co2_mm_mlo.txt.

[32] E. K. Berner, R. A. Berner: Global Environment Water, Air, and Geochemical Cycles. Princeton University Press, 2012.

[33] http://www.esrl.noaa.gov/gmd/ccgg/trends/.

[34] Rudge J. F. et al.: Broad bounds on Earth's accretion and core formation constrained by geochemical models. Nature Geoscience, Volume 3, 439–443 (2010).

[35] Lu, Lu, Zhou et al.: Evidence for direct molecular oxygen production in CO2 photodissociation. Science Vol. 346, Issue 6205, pp. 61-64 , 2014.

[36] Rahmsdorf, S. und Schellnhuber, H.J.: Der Klimawandel. Verlag C.H. Beck, München, 8. Auflage 2018.

[37] Sagan, C. und Mullen, G. : Earth and Mars: Evolution of Atmospheres and Surface Temperatures. Science 177, 1972, S. 52–56.

[38] Etheridge, Etheridge, D. M., L. P. Steele, R. L. Langenfelds, R. J. Francey, J.-M. Barnola, and V. I. Morgan, (1996), Natural and anthropogenic changes in atmospheric CO2 over the last 1000 years from air in Antarctic ice and firn, J. Geophys. Res., 101(D2), 4115–41.

[39] Petit, Petit; J.R. et al.: Climate and atmospheric history of the past 420,000 years from the Vostok ice core, Antarctica.Nature volume 399, pages 429–436 (1999).

[40] Bereiter, Bereiter, B. et al.: Revision of the EPICA Dome C CO2 record from 800 to 600 kyr before present. Geophysical Research Letters, 2015. doi: 10.1002 / 2014GL061957, https://www.ncdc.noaa.gov/paleo-search/study/17975 (01/2020).

[41] D. M. Etheridge, L. P. Steele, R. L. Langenfelds, R. J. Francey, J.-M. Barnola, V. I. Morgan: Natural and anthropogenic changes in atmospheric CO2 over the last 1000 years from air in Antarctic ice and firn. In: Journal of Geophysical Research. 101, Nr. D.

[42] Umweltbundesamt: http://www.umweltbundesamt.de/sites/default/files/medien/370/dokumente/jm_co2.pdf (21.11.2015).

[43] F. u. S. P. Scheffer, Lehrbuch der Bodenkunde. F. Enke Verlag, Stuttgart, 1976.

[44] Stumm, W. und Morgan, J.J.: Aquatic Chemistry. John Wiley & Sons, New York, 1981.

[45] Scheffer, F. und Schachtschabel, P.: Lehrbuch der Bodenkunde. F. Enke Verlag, Stuttgart, 1976.

[46] Grohmann, A.: Physik und Chemie des Wassers, in: Bretschneider, H. u. a. Hrsg., Taschenbuch der Wasserwirtschaft, S.47 - 88, Verlag Paul Parey, Hamburg u. Berlin, 1982.

[47] Deutsche Einheitsverfahren zur Wasser-, Abwasser und Schlammuntersuchung D 8 : Berechnung des gelösten Kohlendioxids, des Carbonat- und des Hydrogencarbonat-Ions. Beuth Verlag, Berlin, 1971.

[48] Deutsche Einheitsverfahren zur Wasser-, Abwasser und Schlammuntersuchung G 1 : Bestimmung der Summe des gelösten Kohlendioxids. Beuth Verlag, Berlin, 1971.

[49] Sontheimer, H., Spindler, P. und Rohmann, U.: Wasserchemie für Ingenieure. ZfGW-Verlag, Frankfurt, 1980.

[50] Rohmann, U.: Grundlagen des "Kalk - Kohlensäure - Gleichgewichts". Wasserchemie für Ingenieure, DVGW-Schriftenreihe Wasser Nr. 205, 6.1 - 6.26, Wirtschafts- und Verlagsgesellschaft, Bonn, 1989.

[51] Spindler, P.: Die Kalklösetendenz von Trinkwasser. Wasserchemie für Ingenieure, DVGW-Schriftenreihe Wasser Nr. 205, 7.1 - 7.56, Wirtschafts- und Verlagsgesellschaft, Bonn, 1989.

[52] DIN 38404 Teil 10: Calcitsättigung eines Wassers, Beuth Verlag, Berlin, 1995.

[53] Kern, D. M.: The Hydration of Carbon Dioxide. J. Chem. Education 37, 14 - 23, 1960.

[54] Stumm, W.: Calcium carbonate depositions at iron surfaces. J. AWWA 48,3,300-310, 1956.

[55] DIN 50930 Teil 1-5: Korrosion metallischer Werkstoffe im Innern von Rohrleitungen, Behältern und Apparaten bei Korrosionsbelastung durch Wässer, Beuth Verlag, Berlin, 1993.

[56] Rohmann, U. und Sontheimer, H.: Nitrat im Grundwasser. DVGW-Forschungsstelle am Engler-Bunte-Institut der Universität Karlsruhe 1985.

[57] Siegenthaler, W. et al.: Lehrbuch der inneren Medizin. Georg Thieme Verlag Stuttgart 1992.

[58] H.-J. Voigt, Hydrogeochemie. Einführung in die Beschaffenheitsentwicklung des Grundwassers. Deutscher Verlag für Grundstoffindustrie, Berlin, 1989.

[59] Schwoerbel, Jürgen: Einführung in die Limnologie. : Fischer, Stuttgart, New York, 1977.

[60] Schlegel, H. G.: Allgemeine Mikrobiologie. Georg Thieme Verlag, Stuttgart, 1992.

[61] „https://www.geysir-andernach.de/fachwissen/funktionsweise-geysir/ (11.01.2021)".

[62] DIN 2000: Zentrale Trinkwasserversorgung, Beuth Verlag, Berlin, 1973.

[63] Eberle, S.H.: Die wasserchemische Berechnung der Kohlensäuregleichgewichte unter Berücksichtigung der Komplexierung von Calcium und Magnesium sowie der Anwesenheit von Phosphat, Ammonium und Borsäure Kernforschungszentrum Karlsruhe, Karlsruhe 1986.

[64] J.M. Edmond, and J.M. Gieskes, Geochim. Cosmochim. Acta, 34, 1261 (1970).

[65] Sigg, L. und Stumm, W.:Aquatische Chemie. Teubner Verlag, Stuttgart, 1991.

[66] DIN 38404 Teil 10: Berechnung der Calcitsättigung eines Wassers. Beuth Verlag, Berlin, 2012.

[67] wateq4f: http://www.phreeplot.org/ppihtml/wateq4f.dat.html (27.09.2016).

[68] Eberle, H. und Donnert, D.:Die Berechnung des pH-Wertes der Calcitsättigung eines Trinkwassers unter Berücksichtigung der Komplexbildung. Z. Wasser-Abwasser-Forsch. 24, 258-268, VCH Weinheim, 1991.

[69] DIN 38409-7: Bestimmung der Säure- und Basekapazität. Beuth Verlag, Berlin, 2005.

[70] Maier, D. und Grohmann, A.: Bestimmung der Ionenstärke natürlicher Wässer aus deren elektrischer Leitfähigkeit. Z. Wasser Abwasser Forschung 10, S. 9-12, 1977.

[71] Gimbel, R.: Grundlagen der Chemie wäßriger Lösungen. DVGW Schriftenreihe W 205: Wasserchemie für Ingenieure, DVGW, Eschborn, 1989.

[72] Hölzel, G.: Messung, Regelung und Einstellung des pH-Wertes im Trinkwasser. Vulkan Verlag, Essen, 1993.

[73] Verordnung über Trinkwasser und über Wasser für Lebensmittelbetriebe (Trinkwasserverordnung - TrinkwV) in der Fassung der Bekanntmachung vom 5. Dezember 1990 (BGBl. 1 S. 2613, ber. 23. Januar 1991, BGBl. 1 S. 227) - BGBl. 111 2126-1-7.

[74] Hütter, L.A.: Wasser und Wasseruntersuchung. Diesterweg Verla, Frankfurt, 1984.

[75] Verordnung über die Qualität von Wasser für den menschlichen Gebrauch (Trinkwasserverordnung - TrinkwV 2001), 21. Mai 2001, (BGBl. I S. 959).

[76] Aurand, K. Hrsg.: Die Trinkwasserverordnung. Erich Schmidt Verlag Berlin 1991.

[77] Meyer, E.: Untersuchungen zum Vorkommen von Asbestfasern in Trinkwasser in der Bundesrepublik Deutschland - und gesundheitliche Bewertung der ergebnisse. GWF Wasser-Abwasser 123, 85-96, 1982.

[78] Verordnung über Trinkwasser und über Wasser für Lebensmittelbetriebe (Trinkwasserverordnung - TrinkwV), 22.05.1986, Bundesgesetzblatt 1986 I Nr. 22, S. 760.

[79] Liste der Aufbereitungsstoffe und Desinfektionsverfahren gemäß § 11 Trinkwasserverordnung 2001; Bundesgesundheitsbl - Gesundheitsforsch - Gesundheitsschutz 2002; 45:827–845.

[80] BÖHLER, E. in: Entsäuerung nach der Trinkwasserverordnung, Informationsveranstaltung DVGW, 6. u. 7. Dez. 1993, Peißen/Halle.

[81] „https://de.statista.com/statistik/daten/studie/1109076/umfrage/verteilung-von-land-und-wasser-auf-der-erdoberflaeche/ (07.10.2022)“.

[82] Küster, W. und Thiel, A: Rechentafeln für die chemische Analytik. De Gruyter Verlag, Berlin, 2008.

[83] Hölzel, G.: Einführung in die Chemie für Ingenieure. C. Hauser Verlag, München, 1992.

[84] Spindler, P.: Die Kalklösetendenz von Trinkwasser. DVGW-Schriftenreihe Wasser Nr 205, DVGW, Eschborn, 1989.

[85] „UNFCCC United Nations Framework Convention on Climate Change: http://unfccc.int/process/transparency-and-reporting/greenhouse-gas-data/greenhouse-gas-data-unfccc/global-warming-potentials (18.12.2020)“.

15 Keyword index